CW00631227

Military Modelling

Gerald Scarborough

Patrick Stephens Ltd
in association with Airfix Products Ltd

© Copyright 1974
Gerald Scarborough and
Patrick Stephens Ltd

All rights reserved

No part of this publication may be
reproduced, stored in a retrieval system
or transmitted, in any form or by any
means, electronic, mechanical,
photocopying, recording or otherwise,
without prior permission of
Patrick Stephens Ltd

First published - October 1974

ISBN 0 85059 177 5

Cover design by Ian Heath

Text set in 8 on 9pt Univers Medium
by Stevenage Typesetters.
Printed and bound by The Garden City
Press Ltd, Letchworth, Herts.
Paper: Factotum Cartridge, 85 gsm.
Published by Patrick Stephens Limited,
Bar Hill, Cambridge, CB3 8EL, in
association with Airfix Products Ltd,
London SW18

Contents

Editor's introduction

This, the third title in the new *Airfix Magazine Guide* series, is a complete introduction to the fast-growing hobby of modelling miniature tanks, armoured cars and trucks from Airfix plastic kits. Military modelling has increased greatly in popularity over recent years, and the introduction of cheap model tank kits into the Airfix range has played no small part in this. Not only do these kitsbuild up into excellent models in their own right, but they can also be used as the basis for numerous conversions into hundreds of alternative vehicles. By this means a large collection of different tanks and other AFVs can be built up, and although some obviously have more potential than others, it is safe to say that there is at least one other model to be made from each kit in the Airfix range. This range itself is also being continuously expanded.

In this book Gerald Scarborough, a well-known military modeller and long-standing contributor to *Airfix Magazine,* explains the basic principles of assembling, detailing and converting plastic military vehicle kits. Although intended for modellers with a little experience (complete beginners should read the same author's *Airfix Magazine Guide 1; Plastic Modelling* first), there is no subject in this book which cannot be tackled by anyone possessing the few essential tools and a modicum of patience. Tools, materials, cutting, shaping and marking out plastic sheet and rod, adhesives, paints and painting, and choosing model subjects are dealt with in the first chapter, and the book then goes on to give numerous practical step-by-step examples of easy (and some more complicated) conversions from Airfix kits. The more advanced techniques of 'scratch-building', where complete models are made from raw materials and only use a few parts from Airfix kits, are beyond the scope of this volume and will be covered in later titles in this series. As in everything else, one must learn to walk before trying to run!

BRUCE QUARRIE

one

Basic tools, materials and techniques

Throughout the pages of this book I will be concentrating on models to the popular scale of 1:76 (4 mm : 1 foot) as this is by far the handiest scale if your aim is for a representative collection of military vehicles. It is also popular for 'wargames', and the Airfix range of kits are of course to this scale. The problem of lack of space to display a collection of large-scale vehicles was my own problem, and indeed you are fortunate if you should have limitless space at your disposal. Cost of kits for conversion is also very relevant as few can afford to expend £2 to £3 on a large-scale kit just to cut it about.

To complement the range of military vehicles there are also the many sets of Airfix figures, and their use alongside, or as tank and vehicle crews, will be discussed as we go along.

Before we get to some of the many conversion possibilities this first chapter will deal generally with tools, materials, painting, display and other considerations that will apply to most models you attempt.

Tools

I find in practice that very few tools are required, in fact only a few more than you would use to make a decent

A fine PzKpfw IV conversion as mentioned in Chapter four.

A Bergepanther recovery vehicle, seen here with a Marder based on a Fujimi kit chassis, is an easy conversion from the basic Airfix Panther kit.

The little Universal Carrier has many conversion possibilities, including this flame-thrower version.

job of construction of any plastic kit. You will find that some will be in constant use while others, although used only occasionally certainly make life easier!

Craft knives are, of course, first on the list–I would not recommend using old razor blades–and these need not be expensive. I have now collected four, one with a curved blade, one hooked, one straight and an old one with a blunt curved blade that I use for applying body putty or any other dirty job. Needless to say they need treating with respect as they are sharp–if not they would be no good for the job–and if used carelessly they will cut deep into you. So do take care–preferably cutting away from yourself.

A good steel engineer's rule is ideal, not only for measuring and marking out parts but as a straight edge to guide your knife while cutting sheet plastic card or scoring in planking or other lines on bodywork etc. A good solid pair of geometry compasses I find invaluable for these scoring jobs, also for opening out holes, impressing rivet detail, 'welding' parts together after heating in a candle flame, and numerous other odd jobs apart from that for which they were designed.

One of my other much-used and handy gadgets is a pin chuck which holds a 1/64 drill for use in drilling out the ends of gun barrels, making pilot holes for opening out with larger drills which can then be twirled between finger and thumb. I have often used this small drill to hold parts for painting just by drilling it in where it will not show after assembly, then with the part impaled you can paint all round and leave to dry.

A needle fitted in the end of a short length of wooden dowel can also be used to 'drill' holes if the needle has been heated in a candle flame, though practice on scrap first as it is not easy to judge the heat correctly. Mention of the candle of course leads me to this useful 'tool'. As already mentioned, the heat from its flame can be used for 'welding',

Two conversion possibilities from the American half-track kit. Full construction details for these are given later in this book.

'drilling', and also melting 'sprue' ready for stretching. This will be discussed later along with materials.

Although probably the most expensive single tool, a razor saw (rough cost about £1) will be required at some time to enable you to work some of the larger thicknesses of plastic sheet and to modify kit parts, so it is as well to invest in one of these as soon as you can afford one. If you have a six inch Junior hacksaw this can be used meanwhile, and though not so good, you can get by for some jobs.

your 'tools' and keep them all together in a handy box. Above all, remember to look after your tools, keep them clean and do not abuse them in use.

Finally you will avoid domestic trouble if you equip yourself with a work board, and all you need for this is an offcut of hardboard some one foot six by a foot. You will find that hardboard is better than wood as it has no grain to put your knife off line when you are trying to cut a straight line.

Materials

To supply the needs of plastic modellers there are now many brands of plastic sheet available like Plastikard, Rikokard, Polycard etc, and a few sheets of different thickness are all that are required. The usual sizes are 0.005, 0.010, 0.020, 0.030, 0.040, and 0.060 inch, commonly referred to as 5 thou (ie five-thousandths of an inch) 10 thou, 20 thou, etc, and as a start I suggest the 10, 20 and 30 thou thicknesses be purchased. The larger thicknesses may be required for some particular jobs but often laminations (ie sandwiched sheets) can be used to build up a thicker section.

One of the many fine conversion possibilities from the Airfix PzKpfw IV kit is the Hummel self-propelled gun.

You won't get far without a pair of tweezers so put them on your shopping list of essentials. Make sure they are not too strong though as you need to pick things up and hold them gently which you cannot do if they are very stiff.

There are also a host of odds and ends that you may already have or can borrow from other members of the family, like manicure scissors, nail clippers, hair grips, small files, a nail file, odd pins, nails, needles, elastic bands and Sellotape-in fact it is best to invest in a large roll of Sellotape as this I find extremely handy for holding parts together while the adhesive dries. Wooden cocktail sticks and a lump of lasticine, paintbrushes and 'wet-and-dry' paper will all have their uses, as will be explained as we go along, so start collecting

Microstrip or other strip plastic is also invaluable and I recommend a packet of assorted as the most useful as this will last a long time. It consists of different width narrow strips which have been

Zugkraftwagen with 3.7 cm Flak 36 comprising a new armoured cab and gun on the Airfix SdKfz 7 half-track chassis, as described later.

cut from sheets of various thicknesses all approximately three inches long. You can also obtain it in packets of all one width and thickness, so if you have a particular application requiring a lot of strip of one particular size then this is the obvious answer. Reference to Slaters, the manufacturers, catalogue or to the stocks in your model shop will give you further details.

Plastic rod is also available from Slaters in lengths of approximately 12 inches and in sizes 30 thou, 40 thou, and 50 thou diameter. Single size packs or an assorted pack can be purchased. A shorter rod is marketed under the Riko label but as the range from all manufacturers is always being changed, check for the latest developments at your local model shop or better still in the review columns and advertisements of modelling magazines like *Airfix Magazine*.

A fairly new product is Plastruct, which you may have seen advertised in the modelling press. An impressive variety of sections of 'H', 'T', angle, rectangular hollow, round tubing and various fitting for architectural models are available but the main drawback is the difficulty in obtaining a satisfactory join using the more common adhesives. A special liquid solvent called Plastic Weld is available from the manufacturers but as it contains chloroform and is a 'harmful' irritant to eyes and lungs, its use by modellers with limited experience cannot be recommended.

New products are always coming on the market so a keen eye on the adverts in *Airfix Magazine* will keep you up to date. A recent one from Modakit, 13 Larchfield Way, Horndean, Hants PO8 9HE, is a Modeller's Pack of ready scored plastic card, designed to represent various widths of planking and also sheets with a canvas texture surface for use in tilts or hoods on vehicles. On some of the models pictured in this book this planking was used, and I found it to be a great time saver as well as giving a very accurate representation. Vacuum moulded kits are also available from Modakit and a stamped addressed envelope will bring the current list of what is available.

One of the biggest sources of materials will be the 'bits' box, which will of course only be acquired as more and more models are built. Whatever the kit you convert there will invariably be some bits you don't use or, in some kits, there are two optional models or variants. Always consign these left-overs to a spares box as some time they will come in useful as a part for another conversion. Get into the habit of collecting all sorts of plastic oddments like pen tops, 'give away' kits in cereal packets, out of Christmas crackers, in fact, any old bit that may sometime prove useful. Keep a stock also of nice round pieces of 'sprue', cast off plastic knitting needles, a few bits of wire, old necklace chains, plastic drinking straws and cocktail sticks. If in doubt keep it!

Working in plastic

Later on in this book, as we deal with specific conversions, I will explain some of the techniques that we can use to work the plastic sheet, strip and rod to produce the shapes we want, but a few general notes will not be out of place now.

Cutting

Most military vehicles are made up of flat faces which simplifies construction greatly, very few having rounded shapes except for those that utilise a cast turret or a partly cast hull. We shall therefore mainly be cutting in straight lines from flat sheet. Plastic card is very accommodating in that the thinner sheets cut neatly and cleanly with a sharp craft knife and to obtain an accurate straight line it is only necessary to use a steel ruler as a guide for the blade. However, do hold the ruler down firmly, keeping your fingers well out of the path of the blade and avoiding the use of too much force otherwise this can distort the edge you are cutting. Thicker sheet can be scored deeply with the knife and then the sheet folded when it will snap neatly along the line. The very thick 40 and 60 thou sheets require use of the razor saw.

When sawing, again avoid too much pressure and use long smooth strokes. Short rapid strokes generate too much heat and the plastic will melt and grip the teeth of the blade.

Shaping

Sanding, or, on occasions, filing, again requires a delicate touch to avoid friction building up too much heat. I find that fine (500 grade) wet-and-dry paper is the best material for all 'sanding' jobs, and if used wet there is little danger of creating too much heat. By wet I do not mean that it should be soaked and dripping with water—just a moist surface is sufficient. I find that I very seldom use a file, preferring to use the wet-and-dry wrapped round a six inch wooden ruler.

Marking out

Of course, before we can cut the shapes required to construct our model we must mark these out on the plastic card. Again the steel ruler can be used with a hard, very sharp, pencil being used to make the mark. I always take the dimensions from the plan and re-draw the shape required direct on the plastic, but possibly you have not yet acquired the necessary draughting skill to do this. In some cases it is the only way, but mostly with flat-sided vehicles the plastic sheet can be lined up under the plan and the shape pricked through with a pin just to leave small indentations in the plastic. These indentations can then be joined up using the pencil and ruler. If you don't wish to spoil your plan with a lot of pinpricks then it can be traced first and this used to make the prick marks onto the plastic. Carbon paper I have never found to be satisfactory as it lacks sufficient accuracy of line. The more

Home comfort for the troops! In the next chapter, Gerald explains how to model this little YMCA tea car, which makes a pleasing change from some of the more 'bloodthirsty' creations and would be a useful adjunct to any rear echelon diorama.

Austin K6 6x4 breakdown gantry based on the Austin cab and chassis in the Emergency Set. Construction of this model is described in Chapter two.

complicated shapes and how to mark them out will be described where required as we go through some specific conversions later, and a few short cuts will also be described.

Panel lines and rivets

To mark out doors and panel lines or body planking on, for example, trucks, you will first have to pencil in their location on the parts. This is actually best done in most cases before the parts are cut from the sheet, to minimise distortion. Once again a gentle touch is required as excessive pressure will certainly distort the parts. To score in lines I use my brass compass point with the edge of the steel ruler as a guide. Held almost vertical to the plastic a groove can be gouged out. On thicker sheet the tip of the razor saw can be utilised, again using the steel ruler as a guide.

Rivet detail can be a bit of a problem as this has to be worked from the reverse side of the sheet and calls for accurate marking out of the part on one side with the position of the rivets on the reverse. The rivets themselves are no problem. Place the marked out sheet, reverse side up, on a sheet of cardboard, and just prick through with the point of the compass again using the ruler as a guide. Practice on scrap first to get the feel of the pressure required and avoid making them too prominent as this will look wrong and distort the plastic. Rivets in 1:76 scale are very small so if you are in doubt about your ability to make a neat job then it's probably better to leave them off.

Use of rod and sprue

If you have never used plastic rod before then I suggest you try out a few bends and curves as practice to get the feel of its properties. Slaters rod is quite hard but has the advantage that it can be bent cold to quite sharp radius bends. You will find that, if bent round a wooden dowel or pencil, neat curves can be formed and they will more or less retain their shape. There is some 'spring' so it is best to make a slightly tighter curve than will eventually be required. Sharp bends can be made round the shaft of the compass point while the tapered handle of a paintbrush will provide a variety of radius curves.

No manufacturer yet produces very thin rod but with a little practice this can fortunately be home-made. All you need is a candle flame and a length of the 'sprue' to which the parts of a kit are fixed in their box. Hold this in both hands above the flame of a candle, no nearer than one inch, turning slowly until the plastic above the flame starts to melt. You will be able to feel it start to give and the surface will look wet. Take away from the flame and pull gently apart, when the molten centre should start to stretch into a long thin rod. Experiment by pulling different lengths and thicknesses until you have mastered the technique, and, incidentally, amassed a stock which can be kept in a spare kit box for use as required.

The candle can also be used to make knobs for gear levers by holding the end of a length of stretched sprue or thin rod up to the side of the flame. The plastic will melt and roll back away from the flame and form a domed end on the rod. Large ones can be cut off and used as sidelights or even headlights on model trucks or tanks, so again practice until you have mastered the method and can produce just the size and shape you want.

Adhesives

Polystyrene cement as sold in tubes at all model shops will be the most familiar adhesive for plastic kits and you have doubtless used it and discovered its good and bad habits. The worst of these is, I think, is its tendency to 'string' all over the place. However, its use is pretty basic to plastic modelling. I always stand my tube on a piece of cardboard to catch the inevitable drips and apply it to the model with a wooden cocktail stick or a sharpened matchstick. Unfortunately plastic card and rod do not stick together well with tube cement, and better results will be obtained using a liquid cement. There are a few brands available and my own preference is for Mek-Pak, a clear liquid adhesive which looks very much like water. The bottle top should be kept on as much as possible as it evaporates very quickly and the smell may be offensive to some members of your household! Prolonged breathing of its 'fumes' should be avoided.

Joins are made by holding the parts together (Sellotape is useful here) and the liquid brushed sparingly into the joint with a fine paintbrush. I keep one especially for this job alone but it does not 'set' on the brush so there is no fear of ruining it. Its action is to just dissolve the edges of the parts to be joined thus 'welding' them together. If by accident you should brush too much into the join any excess should be quickly blown off — but do be careful of the direction. As with any new product or material that you use it is profitable to try them out on scrap first until you learn how they can be used and what they do.

Humbrol Body Putty, Plastic Padding and Squadron Green Putty are all fillers which can be used in cracks, mould marks or dimples. Careful construction will reduce the need for this but they are useful to have by you. The Humbrol Body Putty is rather slow to dry, Plastic Padding has to be mixed from the contents of two tubes but is quick drying and Squadron Green Putty comes ready for use in one tube and dries very quickly. It is however, as the warning on the tube says, an inflammable mixture and you should avoid prolonged or repeated contact with skin or breathing of vapour. Although I am not suggesting you would try eating it, it is extremely poisonous and should be kept out of the way of younger members of the household.

As a matter of routine you should always carefully read and follow any warnings or instructions on adhesives,

PzKpfw III and IV models as described later in the book.

fillers or paints etc, not only for your own good but for the sake of others too.

Paints and painting

Military vehicles, that is tanks, armoured cars and trucks, were usually painted in a simple one-colour finish, occasionally with a random camouflage applied over this in one or more different colours. The finish was almost invariably matt although there are some periods in between the two World Wars and since, when peacetime vehicles had a slight glossy appearance. Similarly, they were often 'bulled up' for displays or parades. In general then all vehicles should be matt finish and you can use either Airfix enamels or Kit number 10 in the Humbrol Range. The six colours included are 8th Army Desert Yellow, Afrika Korps Desert Yellow, Olive Drab, German Panzer Grey, Dark Green and Dark Earth, and this set is a good basic buy. The colours are pretty authentic but I don't like to be too dogmatic about the shades of vehicles in wartime as they were subject to great variation due to effects of fading, weather, wear, different paint mixes, locally applied camouflage and many other reasons.

I would suggest therefore that a tin of matt black and matt white at least be added to the above starter set as this will enable you to vary the shades on different models, or indeed on the same model, so that they do not all look alike. Study of photographs will help though colour photographs can be suspect as they seldom give the true colour as this depends so much on the colour of the sky and the light intensity at the time the photograph was taken. It does help to give depth to a model if undersides and parts normally in shadow are painted in a slightly darker shade than the topsides.

Never use cellulose paints on a plastic kit or model, or you will melt it. Whatever paint you use, be it Airfix, Humbrol, Plaka or Testor, you must stir it up well—a quick whisk round just will not do. My usual method when starting a new tinlet is to first, of course, carefully remove the top and then, using a wooden cocktail stick, stir the pigments from the bottom of the tin and break them up into as small lumps as possible. Replace the lid and then shake vigorously for at least five minutes. To save time and energy, particularly energy, I shake up the three or four that I am going to use for the model all together wrapped tightly in a paint rag for safety. After a real good shake then try another stir to see if the lumps have dissolved completely. If not a further shaking will be necessary. It is no good starting to paint until these have all dissolved otherwise the paint will probably not cover and may even dry with a slight sheen or gloss.

The paints are easy to blend to obtain different shades and a collection of small tin lids should be acquired for this purpose. Transfer the paint from the tinlets to the lid using the cocktail stick that you have been stirring the paint with. If you wipe these sticks on an old rag immediately after use they can be used again and again.

Tanks in particular have many parts that are inaccessible to the paintbrush when the model has been assembled, so it is best to paint as you go. For example, the hull should be painted before the roadwheels, idlers and sprockets are fixed in place and these should be painted separately beforehand. Some modellers like to paint the wheels while still attached to the sprue but I prefer to cut them away and then push a cocktail stick through the axle hole and paint them on this. A lump of plasticine on the workbench is handy to stick them into while the paint dries.

As you will already have made up some kit models you will have a few paint brushes and will already have learned that it is a false economy to buy cheap ones. Just two good brushes are worth a dozen cheap ones as there is nothing worse then one whose hairs stick out at all angles or come out all over the model you are trying to paint. I would suggest a Number One and a Number Two Windsor and Newton squirrel hair as being perfectly adequate for small models like the tanks and vehicles we are going to make. If you have a bit of spare cash get the size smaller as well as it will come in useful for finer detail. The main thing is to look after them well by washing out in thinners immediately after use. Cellulose thinners will do a better job of cleaning

A Crusader tank trying hard to look like a lorry, one of the camouflage tricks used in the desert during the Second World War.

Bedford QL-T personnel carrier model based on one of the vehicles in the Airfix Refuelling Set.

Bedford OY 4x2 GS truck, a simple conversion based on the Austin K6 cab etc. Details on constructing all three of these models can be found in this book.

but it must not be allowed to come into contact with your model as it just dissolves it. This means that the brush must be dried well, and a soft rag used gently will squeeze most of the thinners out and the brush will then soon dry as they evaporate rapidly. For this reason also the lid should be put on the tin as soon as possible. *Never* use cellulose thinners to thin down enamel paints.

There are many books available dealing with camouflage and markings of military vehicles and I would suggest that a visit to your local library would be the first step in increasing your knowledge. They may even have some of the bound editions of the Profiles like Armoured Fighting Vehicles of the World, Vol.3: *British and Commonwealth AFVs 1940-1946* or *Armour in Profile.* Some of the Almark series of books like *British Military Markings, 1939-1945, Wehrmacht Camouflage and Markings, Soviet Combat Tanks,* etc, all contain useful gen on colour schemes. Almarks uniform series are also a help in getting 'crews' correctly dressed.

Use of model

'Use of model' may sound a funny subheading but you must have some reason for wanting to make military vehicles apart from the pleasure of accomplishment. My own fascination simply grew from the first few Airfix kits I made which led originally to a desire to know more about tanks. From photographs and information gleaned from books on the subject I then started to modify the Airfix kits to different models and, though a trifle crude by today's standards, these early models helped develop my modelling techniques so that I tackled more ambitious and more accurate projects as knowledge increased.

Over the last few years of course the amount and quality of published information in books and hobby magazines has grown tremendously so that now newcomers to the hobby can find most of what is required with relative ease. My interest led to a desire for a representative collection of tanks and this in turn to a need for some of the support vehicles, transporters, workshop trucks, command vehicles, general service trucks, fuel carriers, personnel carriers, wreckers and rescue vehicles–the list is endless. My collection is by no means complete of course as there are hundreds of different makes, types and variants used by the combatant nations of the Second World War alone, as well as hundreds more of both before, and after, to say nothing of the prototypes – this leaves me with many years of modelling to look forward to! However, just building up a collection may not appeal to you as an ultimate aim for various reasons such as lack of space or a more specialised interest.

The Second World War can be split into different sections, for example, up to Dunkirk, the North African campaign, the Russian front, the battle for Italy, D-Day, the offensive in Europe and the battles against Japan, to give just some broad lines of demarcation from which to choose. I admit to a bias for the Desert battles but your own preferences will probably be different. May I suggest therefore that a more modest collection could specialise in one smaller area of conflict with the aim being to model each type of vehicle used at that time. Any collection should aim to contain individual models that, if you like, 'tell a story' by which I mean they should be detailed and equipped, probably crewed, so that they convey a little of the atmosphere of the time. This is where your study of photographs of vehicles in action can be of such a help, and indeed is essential. There are some modellers who have the knack of creating a model that is instantly recognisable although it may in fact not be dimensionally absolutely accurate to the last millimetre. This atmosphere is, I think, more important than accuracy so don't worry too much if at first you are unable to model to fine limits. If you get to within half a millimetre you are pretty good.

Use of models in a little diorama or scene is now a very popular form of modelling vehicles and this is taking the 'telling a story' idea a stage further. The base and scenery of course complement the finish and detailing of the vehicles or group but require careful planning to obtain the best results. For your early attempts I would suggest you try to recreate some photograph or a part of a photograph in model form. Of course pick something simple as this will not only be easier but probably a lot more effective. Study of other modellers' ideas as displayed in exhibitions or illustrated in magazines will show some of the do's and dont's. For instance, to my eye many dioramas often look far too cluttered with vehicles and troops crammed in so that the density is something like Wembley on Cup Final day! A simple diorama like a tank crew having a 'brew up' or refuelling or re-arming from a truck can be far more effective and easier on the eye. Imagination and observation are probably the key though accuracy and modelling skill are important for success.

If you are not recreating a scene from a photograph then try to sketch out what you are trying to portray and set the models out on the table, varying their positions for the best effect from all angles. A sketch plan can then be drawn up for reference when you start to make the permanent display base.

By entering some of your own models in competitions you will be able to compare your standards with those of other enthusiasts and then by analysis and self-criticism you will know where to improve your efforts. You may of course even win!

Wargames vehicles require a slightly different approach to the detailed models that you would make for a collection or use in a diorama. Although basic construction principles will be the same the aim should be for something a lot stronger to cope with the constant handling. Some deviation from scale accuracy is therefore permissible and thicker plastic card should be used for things like truck bodies and no attempt need be made to add a lot of detail fittings, as these will invariably get knocked off with the constant movement of 'battle'. Make certain that the axles and wheels are firmly attached, strengthening where this will not show by adding additional gussets of scrap plastic. There is nothing worse than your transport going unserviceable just when you are ready to make the decisive attack.

Choice of prototype

The choice of vehicle to model will depend on the use as discussed previously, but whatever it may be you will need a drawing from which to work. There are now several sources from which these can be purchased apart from those often featured in *Airfix Magazine* and other similar publications. Some of the most popular are those by Bellona, now published by MAP Ltd, and a check through their adverts will show what is available. Others are produced by John Church, 'Honeywood', Middle Road, Tiptoe, Nr. Lymington, Hants, and by Len Morgan, 34 Midfield Court, Thurtlands II, Northampton and these are perfectly adequate and relatively inexpensive. A stamped addressed envelope will bring you their complete lists of what they can offer.

There is also a Miniature Armoured Fighting Vehicle Association which publishes a bi-monthly magazine, *Tankette,* which contains many useful drawings, photographs hints and tips, kit and book reviews etc. A line to the Secretary G.E.G. Williams, 15 Berwick Avenue, Heaton Mersey, Stockport, Cheshire SK4 3AA will bring you further details. An incidental benefit of MAFVA membership is that members only can purchase kits and parts for conversions from Mr Eric Clark. These are produced as fibreglass mouldings and, though simple, they do incorporate an amazing amount of surface detail.

The most satisfying models, I think, are those which you have made from your own drawings and probably, as you get more experienced, you will want to attempt this. There are still a lot of Second World War army vehicles either in use as garage wreckers, by fairground operators or just derelict in scrap yards etc, so always be on the look-out and, after obtaining permission, try to measure them up, roughly sketch them and photograph from as many angles as possible.

The Imperial War Museum, London, and the Royal Armoured Corps Tank Museum at Bovington, Dorset, are of course the places to visit if at all possible as there you can examine the real thing in detail. There may be local museums that have a few exhibits of a military nature but these mostly tend to be uniforms or weapons although you may come across some interesting photographs or paintings. Photographs are obtainable from the Imperial War Museum and the RAC Tank Museum publish some inexpensive books, photographs and some drawings although most of these are to scales larger than 1:76 which have to be scaled down.

two

Military transport vehicles

It is perhaps surprising that kit manufacturers have so far paid little attention to the many hundreds of types of support vehicles used by the combatant nations in the Second World War.

In this chapter then I hope to go into some conversion ideas using the Airfix Airfield Service sets as a basis. In a book of this size I cannot hope to cover them all but I hope the examples chosen will give you some ideas and explain some of the methods that can be used.

Models from the Emergency Set

The first of the Airfix releases was the Emergency Set consisting of an Austin K2 Ambulance and an Austin K6 Fire Tender. The K6 was a fortunate choice as this chassis was used as a basis for many specialist bodies like signals, breakdown, office, stores etc, as well as the general service type body, and this same chassis, with just an alteration to the correct wheelbase, will serve for Albion, Leyland, Thorneycroft, Guy, Karrier and AEC types.

Austin K3/YF

Just to get you in the mood I would suggest for a first attempt the Austin K3/YF Truck, 3 ton, 6x4 GS. For beginners the '6x4' nomenclature relates to the wheel arrangement, that is it has six wheels of which four are driven. This is a universal method of description. Some other arrangements are as follows: 4x2 (four wheeler, two driven), 4x4 (four wheels, all driven), 6x6 (six wheels, all driven), 8x6 (eight wheels, six driven) and 6x4-8 (six wheel tractor unit with four wheels driven fitted with eight wheel semi-trailer, eg Airfix Scammel Tank Transporter).

But back to the K3/YF GS (General Service) Truck which is a combination of the two vehicles in the Emergency Set as it uses the K6 chassis with the K2 bonnet and mudguards all combined with a scratch-built body.

To modify the cab first remove the windscreen and the folded side curtains (Part No A37) and assemble the bonnet top, sides and front to this cab front. Cut off the cab floor from Part A1, just behind where the floor steps up in level. Use the razor saw for this job and you'll find it easier to turn the part upside down with the blade up against the step in level. Always remember to clean up saw cuts with a fine file or wet-and-dry paper. The mudguards (part A3) floor and bonnet/cab front can now all be cemented together. Cut a new cab back and top with 'hood' from tissue paper as shown in the sketch, taking your measurements from the scale drawing

The author's Austin K3/YF model.

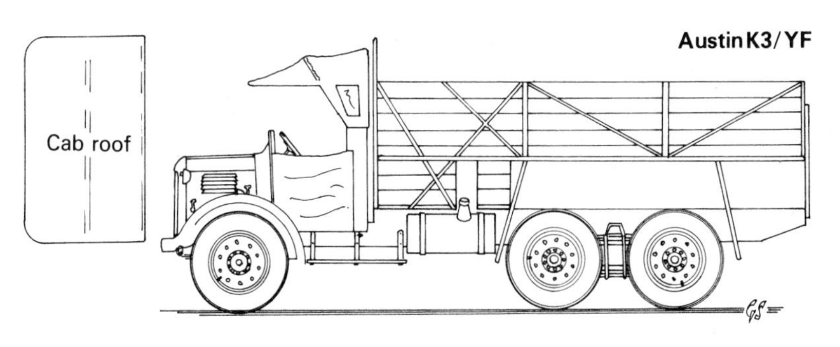

1:76 scale

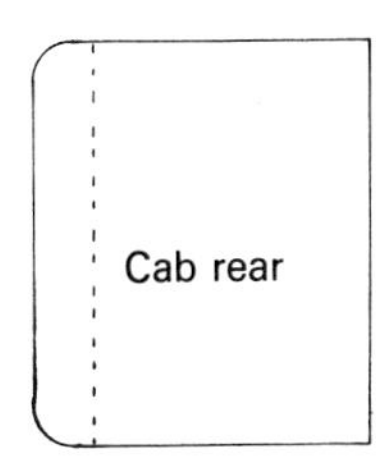

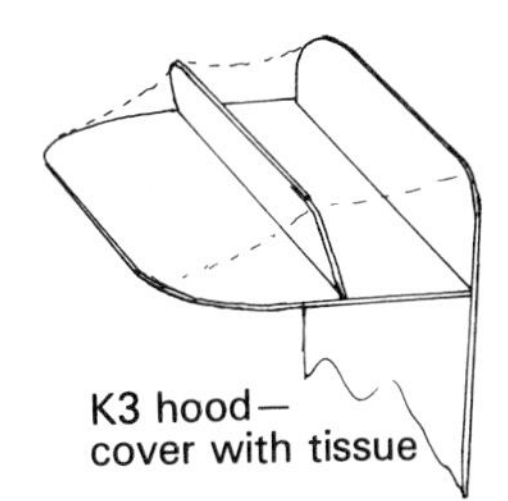

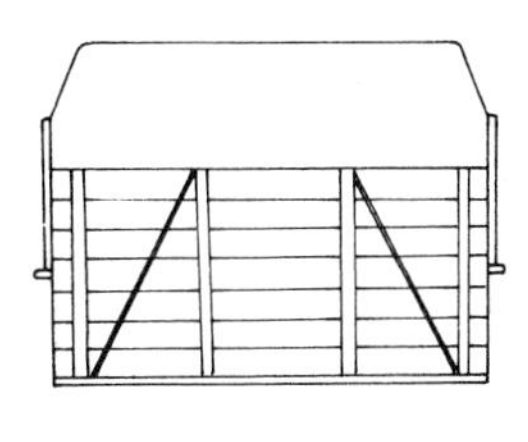

and fitting it vertically to the cab floor. The narrow side strips can then be added and the inside painted ready for fitting the 'canvas' side protection 'doors'. Again tissue paper can be used for these, or Modakit's canvas texture plastic card.

The K6 chassis can now claim our attention, giving the cab assembly time to dry out thoroughly. Always try to plan your construction methods so that your models are built in small sub-assemblies which can all be cemented together at the end as this does allow time for them to dry. The chassis (part No 3) should have the fuel tank removed and approximately 1.5 mm shaved from across the front to allow the K3 cab to seat properly. A larger fuel tank, to the dimensions shown in the drawing, is made from plastic card, 15 or 20 thou will be fine, made up into a 'box' with the corners rounded off on the outside corners at least and with straps from thin strips of Microstrip. The K3 front axle (part A13) is used, adjusting to fit the K6 springs. All springs and axles can be fitted making the distance between rear bogie centres 16 mm. Just slip the wheels in place on the axles and check that they all touch the ground, although this point does not necessarily apply if you intend your model to be part of a diorama where the suspension is re-set to suit the particular ground contours of the base.

For the well-type body I used the Modakit 'planked' plastic sheet but the alternative of course is to score in your own plank join marks by using a compass point as described in the first chapter. Mark out the sides, bottom, front board and taildoor, cutting these out accurately and checking that they will fit together before cementing in place. I hold all the parts together with

K3/YF GS truck body assembled from planked plastic sheet.

small strips of Sellotape and then brush Mek-Pak sparingly into the joints, making sure of course that the whole assembly is square. Make a point of regularly doing a test fit or 'dry run' of all parts before cementing into place as this does give you a chance to adjust for fit if required.

When all the painting has been done the wheels, which should also have been painted separately, can be fitted in place. You will see on this model, and on some of my others illustrated in this book, that the front wheels have been cemented at an angle to the direction of travel to give the impression of the vehicle turning.

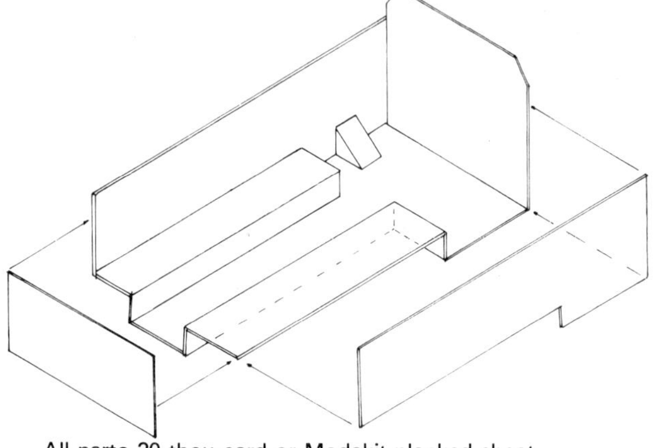

All parts 20 thou card or Modakit planked sheet

All left-over parts from this and other conversions should be kept in a 'bits' box or filed away in their original boxes so that they are available for other conversions. The left-over Austin K6 cab etc can be used to make a Bedford OY GS Truck.

When the basic body is completed to the stage shown in the photographs the detail can be added on the sides and tail door. This is from Microstrip which can be fitted oversize and then trimmed off flush with the top when dry.

Before fitting the completed body to the chassis it is as well to get some of the underside painting at least done, paying particular attention to the parts that will be inaccessible later. After a final check for fit the body can be cemented in place and when dry the painting can be completed. You may find it easier to leave off the rear mudguards until after the body has been fitted but for illustration purposes I have shown them in place.

Another view of the author's completed Austin K3/YF truck model.

Bedford OY 4x2 GS truck

The drawings for this and the sketches were originally published in *Airfix Magazine Annual 1973* but are repeated here as it serves to show how spares can be used up. Before the introduction of the Bedford Tractor Unit and Queen Mary Semi-trailer in the Airfield Recovery Set, I and many other modellers had used the Austin K6 to model Bedford variants and I still think that this is the most economical way to make a Bedford. I will mention what I consider to be some of the shortcomings of the Recovery Set Bedford later. To return to the OY types, the K6 cab (parts 28 to 33) with the hip ring in the roof removed and filled, are assembled, painting inside, and then the bonnet is cut off with a razor saw. The seats are added to the

Completed Bedford OY 4x2 GS truck model in 'desert' setting.

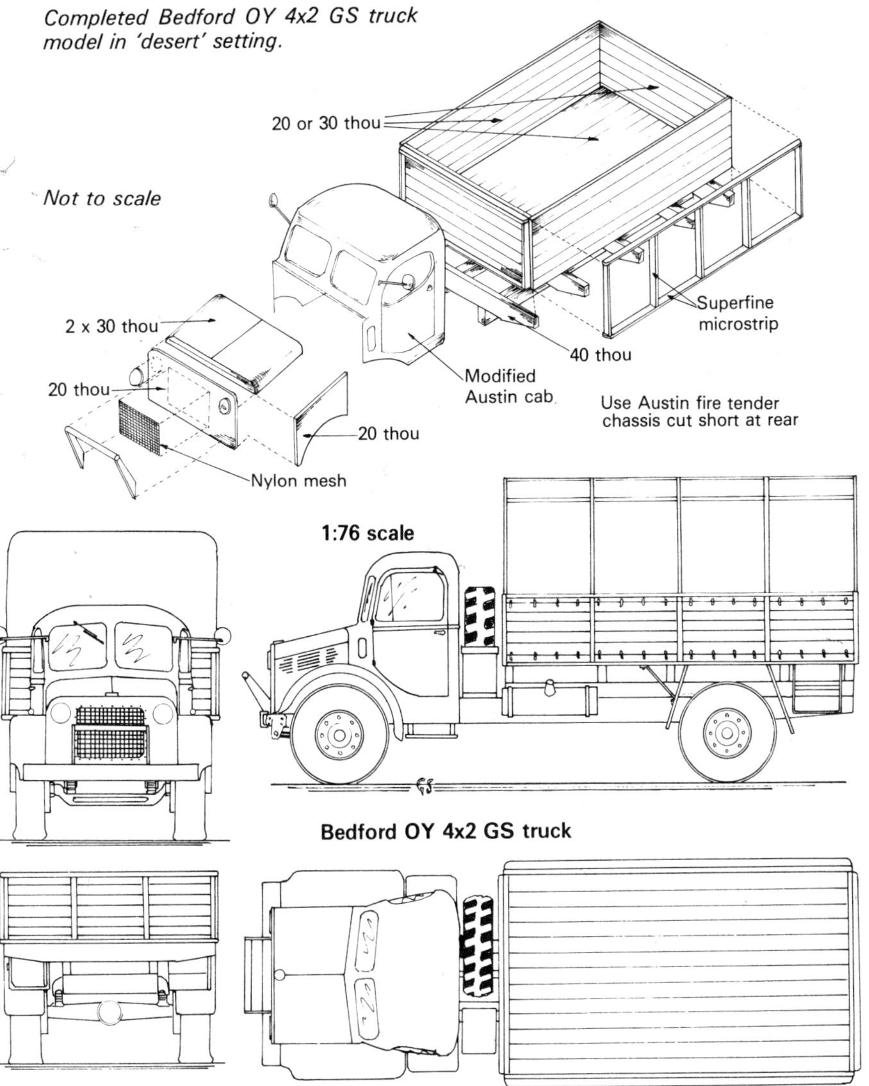

floor and the mudguard edges rounded off. Plastic card is used for the new bonnet with a square of nylon stocking stuck on for the grille. The chassis can be built from sheet using the ambulance springs and axles left over from the K3/YF conversion or made from plastic rod or plastic cocktail sticks. Note some of the extra details like mirrors etc fitted to my model.

Austin K6 6x4 breakdown gantry

An interesting series of breakdown and gantry vehicles were used during the Second World War and the one that I have chosen is built on the Austin K6 chassis. The cab and chassis are virtually as from the kit, the only modifications required being an extra fuel tank and step on the offside to match those on the nearside. These can either come from the spares box or be made up from plastic card. The hip ring can be covered with a tissue disc if required. Shorten the end of the chassis to the length shown on the drawing and add the new crosspiece across the rear with a spring hook from a Matador or similar. The well-type body is made in the same way as that which we made for the K3/YF, but is actually simpler. Details are shown on the sketch and the photographs.

The tilt hoops and gantry supports are the next stage and the easiest way is to construct these over a sheet of greaseproof paper laid on the drawings. Wide Microstrip is used, building up the shape with the two uprights, the top, and the corner gussets. Note the rear one goes

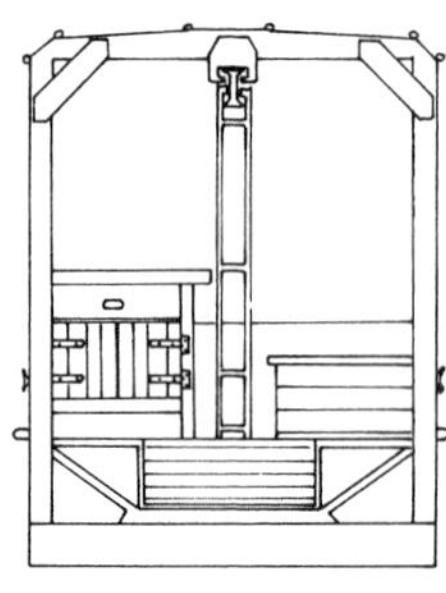

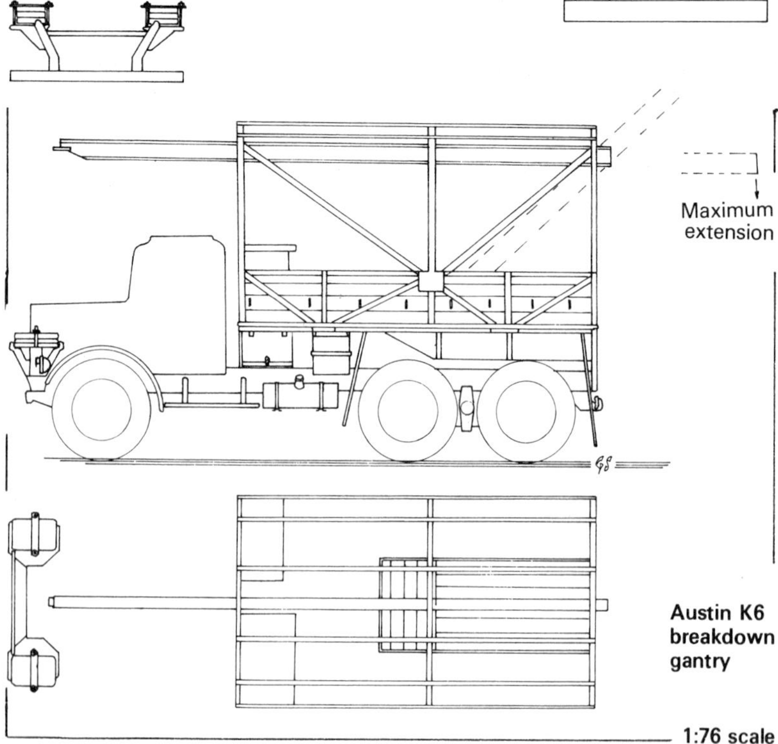

Austin K6 breakdown gantry

Austin K6 breakdown gantry chassis. Note hook from Matador at rear.

Completed model awaiting final painting. Note in this photo the jib is stowed in the travelling position: a view with it in the lifting position appears on the next page.

right down to the crossmember that was added to the chassis. When dry they should be added to the body, taking particular care that they are square and vertical, and then the diagonal braces cemented in place to the body sides and gantry supports. The gantry beam is from H-section Plastruct

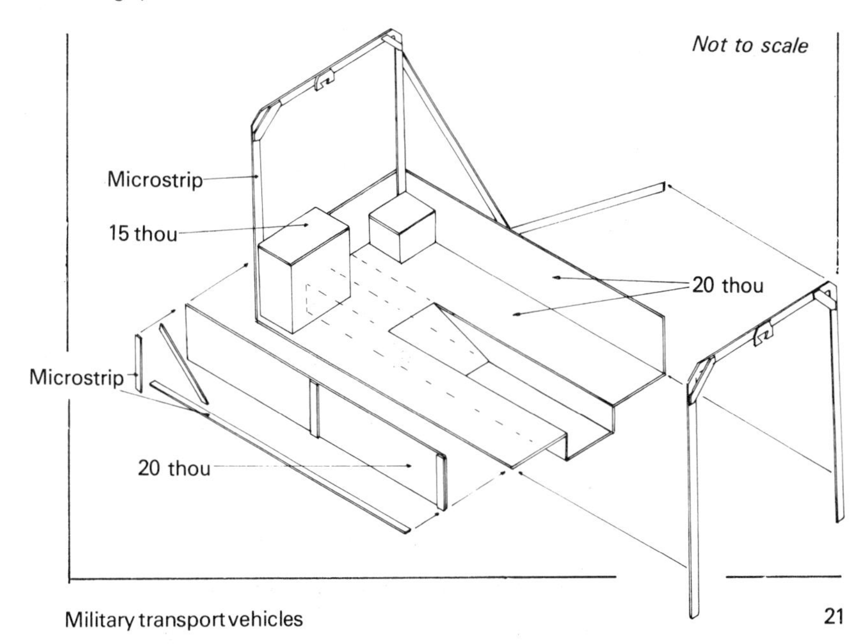

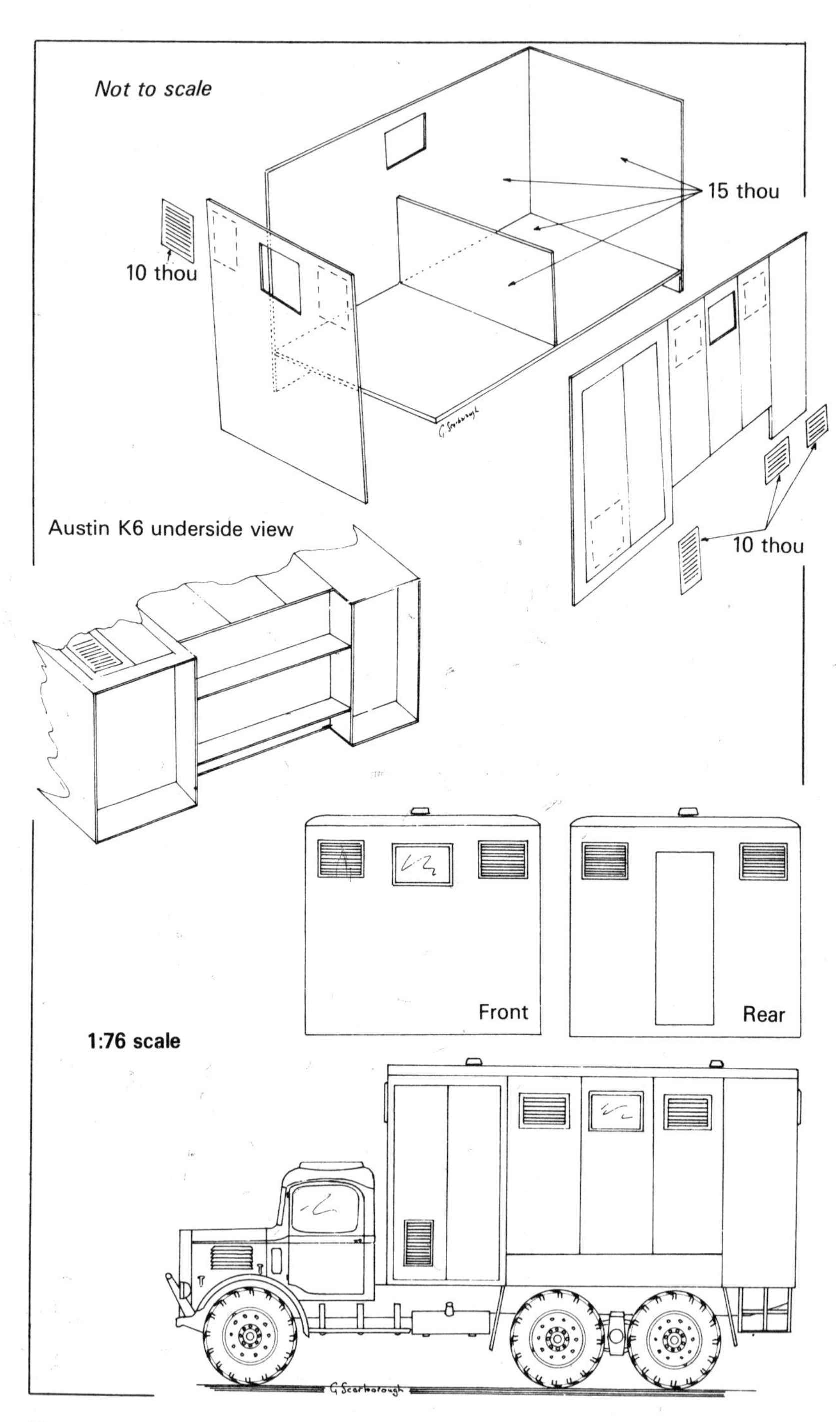

Not to scale
15 thou
10 thou
10 thou
Austin K6 underside view
Front
Rear
1:76 scale

Another view of the completed K6 breakdown gantry showing jib in the lifting position: compare with photo on page 21.

A simplified K6 breakdown gantry model suitable for wargames where robust construction and ease of assembly is more important than small detail.

Austin K6/ZB signals/wireless van which can be made using the plans and construction drawings on the facing page.

and can be made to slide in brackets fixed to the top. Photographs again show this beam in various positions. The longitudinal tilt support rods are added to complete ready for painting and assembly to the chassis.

At the front of the chassis the ballast weight frame and weights are constructed from thick plastic sheet as the sketch. A simpler model can be made if the tilt is modelled in place and I have included a photograph of an old model of mine to show this. Several inaccuracies will be apparent in this old model but at the time it was made for wargames where the simplified construction was acceptable.

Austin K6/ZB signals/wireless van

The drawings and sketches which originally appeared in *Airfix Magazine* in 1970 have been repeated here to show yet another simple conversion on this chassis. A look through the list of John Church drawings will show other Austin K6, Austin K2 and Bedford variants that you may like to try, using the methods described.

RAF Refuelling Set

The Bedford QL types are probably the longest serving of Second World War vehicles, and in fact some are still to be seen, mostly in agricultural applications, in use in 1974. The slatted seats of the QL-T Troop Carrier left a lasting impression on many servicemen but 'drooper', as it was popularly known, is probably remembered with affection for its reliability and anyway it was better than walking!

The pressed steel-type body is a little tedious to reproduce but by making up the plain body first this detailing can be added from thin Microstrip on the outside panels with the uprights and top edges also from this material. Add the slatted seats before the tilt frames, and it is also a good idea to paint as you go along. Note the seats along the sides do not come right up to the front as there are doors here to assist rapid exit.

The chassis is straightforward, just follow the kit instructions, parts 1 and 2.

Bedford QL-T

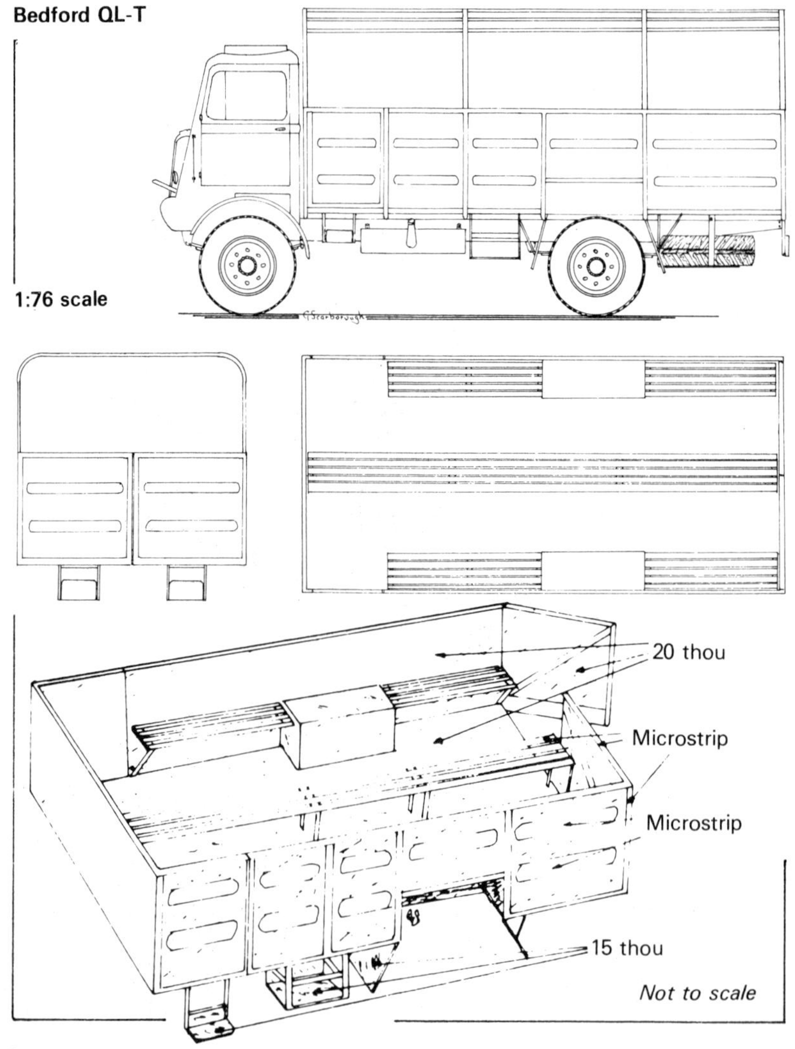

Author's Bedford QL-T troop carrier based on the chassis of the Bedford QL in the Refuelling Set.

If you have a spare steering wheel from another kit this can be used inside as for some reason this has been omitted from the kit. The front mudguards can also be carved away on the inside so that they appear to be of thinner material, and a door can also be left open if desired. The rear chassis should be extended to the length shown on the drawings before the body is fitted and then the steps, fuel tank, water can and spare wheel carriers added.

Other QL variants are also in the John Church list and to be found in the Bellona Military Vehicle Data Series.

AEC Matador 6x6

Unfortunately this chassis was not used for many other purposes so you are likely to have one of these in your spares box until such time as you do some more complicated scratch-built projects. Here the wheels, axles, etc, will be of use for some of the Albion, Leyland and Foden 'heavies' so keep these until you have more experience.

One that is within the scope of the beginner is the crane which uses a similar Coles Mk VII to that on the Thorneycroft in the Recovery Set. Construct the AEC chassis as the kit instructions but leave off the rear mudguards and of course the wheels at this stage. Add strips of 60 thou plastic card to the top of the chassis frame from behind the cab to the rear to bring up to the depth as shown on the drawing, filling the mudguard locations with scrap or body filler. The platforms for the auxiliary engine and generator housing and the crane turntable are cut to the same width as the cab and are from 40 or 60 thou plastic card. Body bearers for the rear platform are also from thick plastic card but note those at the mudguard position are not full width. The AEC rear mudguards with their locating brackets removed can be cemented in place under the platform. The turntable for the crane can be cut from the Thorneycroft and used as shown on the photograph. The generator housing is built up as a box of plastic card or can be based on rectangular hollow section Plastruct. The crane part is straight from the Recovery Set, and

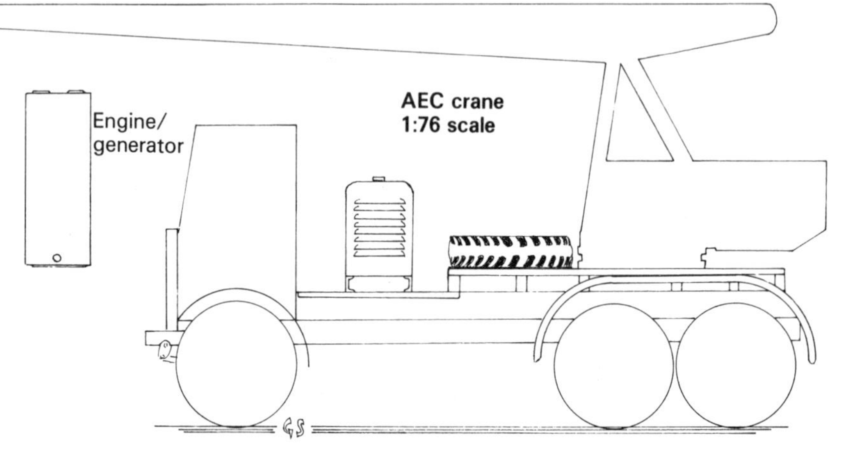

though my model is very plain, being put together just for the purposes of this book, it can be detailed with mirrors, grilles to the generator housing, and spot lights on the roof, etc, together with the usual litter of chains, slings and cables to be found on vehicles of this type.

Recovery Set

As previously mentioned, modellers had used the Austin K6 as a basis for the Bedford family of army trucks until the introduction of the Recovery Set. This Bedford, in the opinion of a lot of modellers, is too small for the Bedford OY types, especially around the front mudguard arches. This can be improved somewhat by carving away the inside of the wheel arches but it is better to use those from an Austin or make new ones from plastic card. This can be done by cutting a strip wide and long enough for each mudguard from approximately 20 thou sheet. This is then curved round a length of wooden dowel of a diameter comparable with the inside shape of the mudguard. The strip is held in place with masking tape (Sellotape will *not* do for this job) and immersed in a cup or bowl of boiling water for about 30 seconds. The strip is then held in cold water to set the shape and then, after removing the masking tape, it can be slid off the dowel. Allow to dry then shape the edges to suit. This method of making up mudguards will be found extremely useful for many models of trucks and I suggest you practice the moulding technique with a few scrap strips until you get to judge the correct length of time required for different thicknesses of plastic.

Another conversion obtained by combining parts from two kits is the Bedford

Matador Crane lorry with new platform body fitted, engine/ generator housing and crane mounting.

General view of the author's completed Matador crane conversion, which could be made more interesting by the addition of small detail items as suggested in the text.

tanker shown in the photographs and drawings. This as you will be able to see uses the modified Recovery Set Bedford cab fitted with larger radius mudguards, combined with the rear chassis from a Bedford QL refueller from the Refuelling Set. The tank is made from two Refuelling Set tanks and utilises three pairs of the bearers (parts 31-34). The tank parts from the two kits (parts 42,49 and 51) are cemented together then a length equal to half the length of the tank required cut from the front of each. These are then fitted, cut ends together, to produce one complete oval tank. Bands of thin (10 thou) plastic card (or even writing paper will do) are fitted all round the tank at front, centre and rear. The other fittings like steps and walkways, fuel tanks, rear mudguards, etc, can come from the spares box, mostly from the rear of the Bedford QL refueller. Just one other point on the wheels of the Bedford OX tractor unit, these are too narrow and can be improved considerably by inserting a section of plastic card between each set of halves. The easiest way to do this is to file the inside faces of the wheel halves flat and cement them down on a sheet of 15 thou plastic card. When dry cut round the wheel and trim up with file or sandpaper and then cement the other half wheel so that the plastic sheet is sandwiched. They also look better cemented on to the axles back to front so that the deepest 'dish' is to the outside.

A natural conversion for the Bedford OX Tractor unit is to replace the Queen Mary semi-trailer with either GS Flat Platform or Petrol Tanker semi-trailer. Again I think the photographs of my

Bedford QL cab - note inside mudguards have been carved away to give an illusion of thinner metal.

Above *completed Bedford refueller: note the Austin mudguards used to replace those provided in the kit.* **Left** *two views of the model under construction clearly showing how the parts from the two kits are joined together.*

1:76 scale

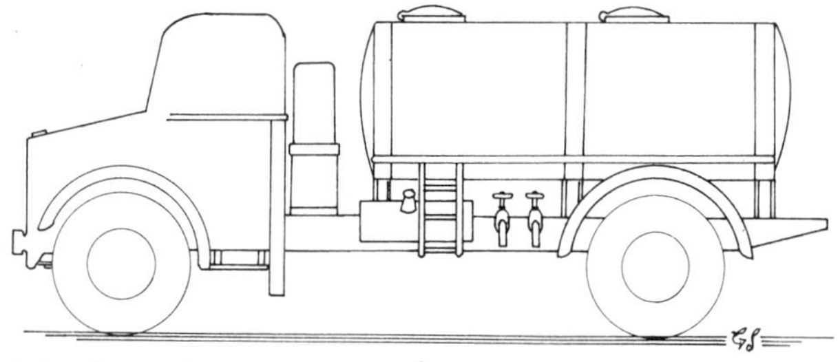

model will shown what is entailed and the drawings show dimensions. A tilt frame could be fitted if required.

A Petrol Tanker Semi-trailer can be made using the same methods we used for the tanker conversion described earlier, and drawings for this are available from John Church's list.

As a result of a letter from another enthusiast, I was spurred on to complete a languishing project to produce some type of refreshment or NAAFI type truck. The 'tea van' must have been a most welcome sight to servicemen and its inclusion in an airfield scene or as part of a collection of representative vehicles is almost essential, although it is not strictly a military vehicle.

The one I have drawn is based on the

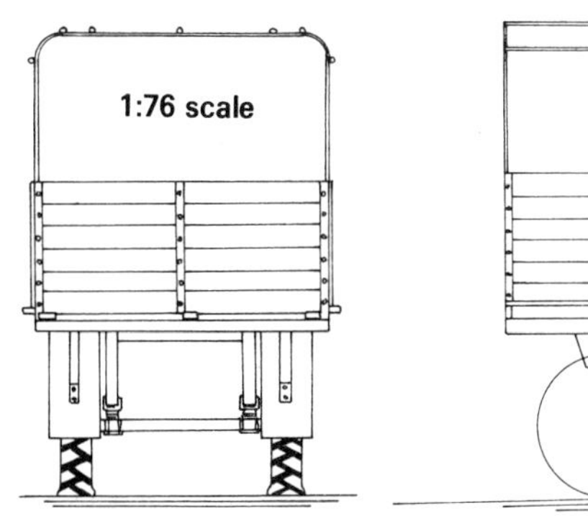

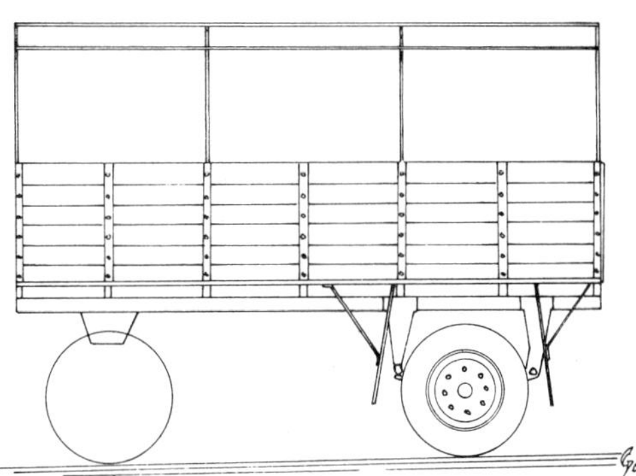

Right *GS semi trailer for use in place of the Queen Mary aircraft trailer shown under construction.* **Above** *the complete articulated unit.*

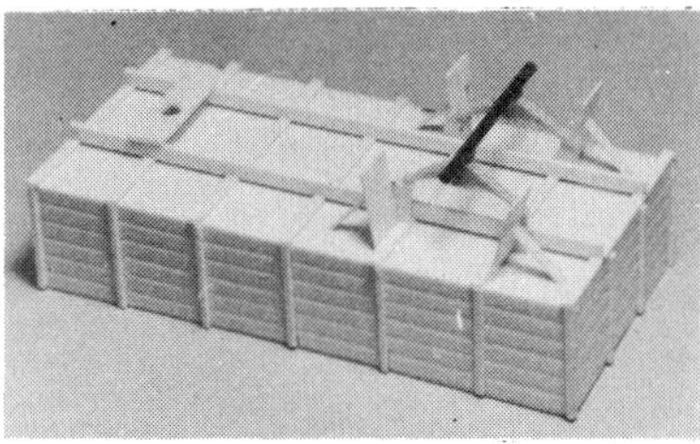

15cwt Bedford for which the Queen Mary tractor unit can be used or for which the cab and chassis can be scratchbuilt as previously discussed. The chassis should be slightly shortened as will be seen in the photograph of the model in construction. Only the bonnet, mudguards and interior detail are used, the rest of the cab being kept in the

Shorten chassis and cut down cab of Queen Mary tractor unit for tea car.

spares box. Refer to the sketch to see how the cab sides are constructed and fit a strip of scrap in to fill the gap on the floor behind the seats. As the van body is a simple square shape its construction should present no problems and I cut all the top, rear, floor, front and roof in one strip, cutting off parts to length using a set-square to make sure they were true. The sides were cut roughly to shape, positioning them so that the wheel axles were correct and then any surplus was cut off all round. The panelling is added using Microstrip but of course if you

Left *the tea car body being built on the modified chassis.* **Right** *the finished model which could also be decorated with miniature advertisements or posters cut from period magazines.* **Below** *1:76 scale plans.*

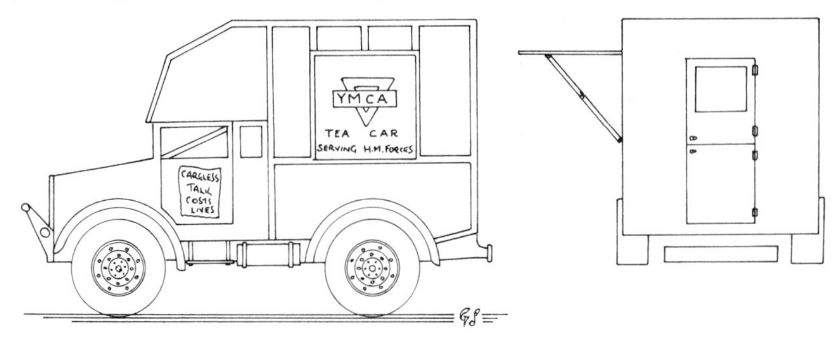

want the serving hatch hinged open you will have to cut out the aperture and fit a counter inside with, say, a tea-urn and a row of mugs from spare sprue etc. My model is in an overall khaki finish and the transfers, a red triangle (which unfortunately doesn't show up too well on the photographs) black stripe with white 'YMCA'and black 'TEA CAR', came from the Airfix 1:24 scale Spitfire kit, although anything similar will do.

Other bodies that could be fitted on a similar chassis are the 15cwt GS truck, water tank or the early type infantry truck without tilt. This latter actually has slightly different bonnet side panels.

Scammel Tank Transporter kit

The Scammel company had a long history and experience of heavy haulage before the Second World War and their vehicles had then, and have retained ever since, a reputation for 'guts'. Ex-Second World War Heavy Breakdown Scammels are still to be seen in use by commercial vehicle repairers going about their recovery jobs as they did during the war.

The Heavy Breakdown 10ton 6x4 model is the one I chose to model for this book as it is typical of the type and I hadn't previously made one to add to my collection.

Before you start construction it should by now be almost routine to study the drawings, sketches and photographs of my model, together with any photographs of the prototype you may have in books. As mentioned you may be fortunate to have a Scammel

Scammel cab under construction: note ballast weights added to front.

Right *Scammel cab back modifications and underside view of basic body.* **Below right** *jib detail and winch installed.*

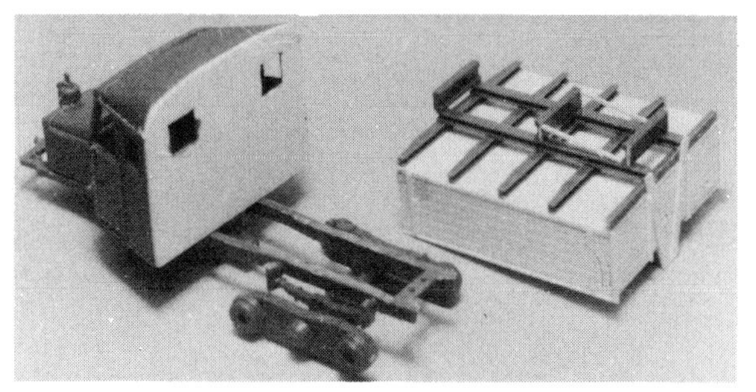

in service still, at a nearby garage, to which you can refer for extra detail, although be wary of any later 'civilian' modifications. Current registration, lighting and Department of the Environment regulations may have enforced detail changes in equipment, almost certainly to the lights, if nothing else.

You will see that there are not many alterations to the kit parts except those entailed in removing the rear crew compartment and shortening the chassis. The floor, sides and top of the cab (parts 5,6,7 and 9) are cut behind the driver's seat and the corresponding join lines across the top and to the rear of the doors. The chassis is shortened by 9 mm and this is best done at a point where the chassis will be cemented below the cab floor. The propeller shaft on the drive train (part 21) will of course have to be

Heavy breakdown 10 ton 6x4

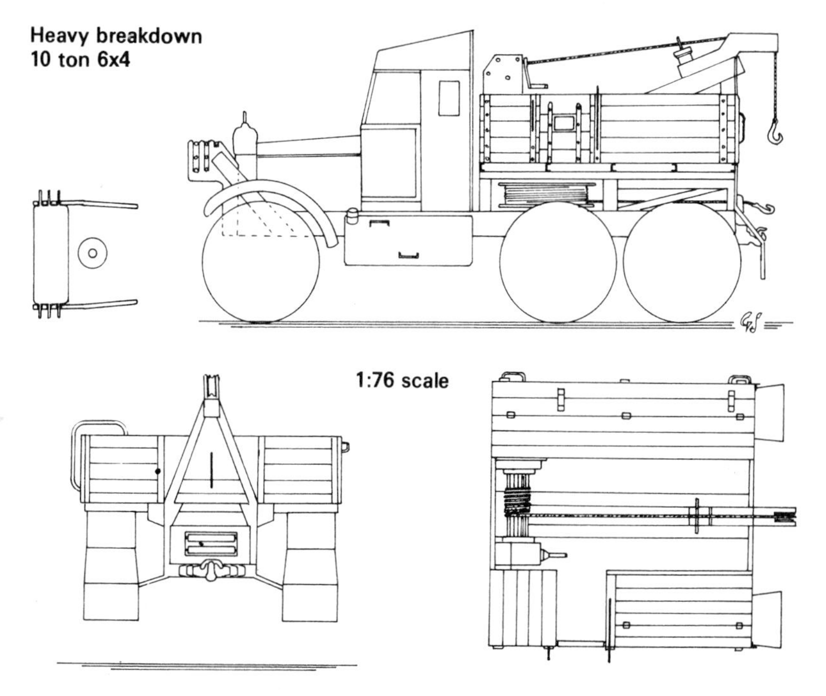

1:76 scale

Scammel model finished in dark brown/green camouflage.

shortened to fit at a later stage. The modified cab and the bonnet sections as in Section One of the instructions can now be cemented together, painting inside of course, and a new back from 15 or 20 thou plastic card cemented in place to the cab. Cut off the top portion of the chassis to which the trailer towing table (part 50) would be fitted so that the top is level. The rest of the chassis parts, suspension, etc, can be fitted at any stage you feel convenient, for example while waiting for other parts to dry. Construction of the rear body follows the methods we have already discussed, first building up the basic 'boxes' to the floor then adding the 'iron-work' from Microstrip. I used Plastruct sections for the body cross bearers with the same material for the sub-chassis and the short cross spacers. The jib and jib supports are again from Plastruct, and 'H' section forming the main jib which slides between a sandwich of channel sections. The winch detail should be visible on the various photographs of my model in construction.

Most wrecker and recovery vehicles were fitted with a ballast weight frame at the extreme front and the Scammel is no exception. On my model this was constructed mostly from scrap plastic to the dimensions shown on the scale drawing.

The body can now be assembled to the chassis fiting the longitudinal winch drum in place at the front, although this is not essential as it is practically invisible when the model is complete. The cable from this winch runs out between two rollers across the rear end of the chassis. All the handles, mirrors, winch wires and hooks, mudguards and stowed drawbars can be added now the basic construction is finished and then, when dry, the complete model can be painted. Wheels of course are painted separately and added last. Crew figures can be found in the various Airfield Service sets and ideas for incorporating the Scammel in a little recovery diorama should come readily to mind.

Long Range Desert Group jeeps are easy to build using the model supplied in the Airfix Buffalo kit plus additional figures.

three

Armoured cars and half-tracks

The Second World War saw the development of the half-track vehicle, a type which had been practically unknown in peacetime except for some very specialised vehicles like the Citroen Kegresse types developed for use in the desert etc. The M3 and its relations used a development of the Kegresse type track assembly used first in truck conversions, later mating it to the Scout Car M3 to produce the familiar M3 Personnel Carrier of the Second World War.

The Airfix kit is generally accurate in outline but appears unfortunately to be a bit of a mixture. Originally the half-tracks were built by Autocar, White and Diamond T, but to step up production to meet the demands of not only the US Army but Canada, Great Britain and Russia, production commenced at the International Harvester factory. These International vehicles were fitted with squarer mudguards as on the Airfix kit but with a hull body of welded construction which had rounded rear corners, not square, as on the kit. There were also minor differences of dashboard and front axle etc, but these are not really relevant in 1:76 scale.

It is obvious then that conversion to the International M5 is simple as this entails merely rounding off the rear corners. To improve the model, however, the interior detail can be modified and I hope the photographs of my model will show clearly what is entailed. The only kit part to modify is the partition between the driver's and the rear compartment (part No 12), which should be cut away as shown in the sketch. The complete vehicle can then be assembled as the kit instructions, painting parts before cementing in place where necessary. Next cut a rectangle to fit to the floor as a mount for the machine gun pedestal. The fuel tank (parts 18 and 19) will have been fitted and you will see that from the rear of

Puma armoured car converted from the basic Airfix kit as described later.

This view clearly shows the new interior detailing required for the M5 half-track.

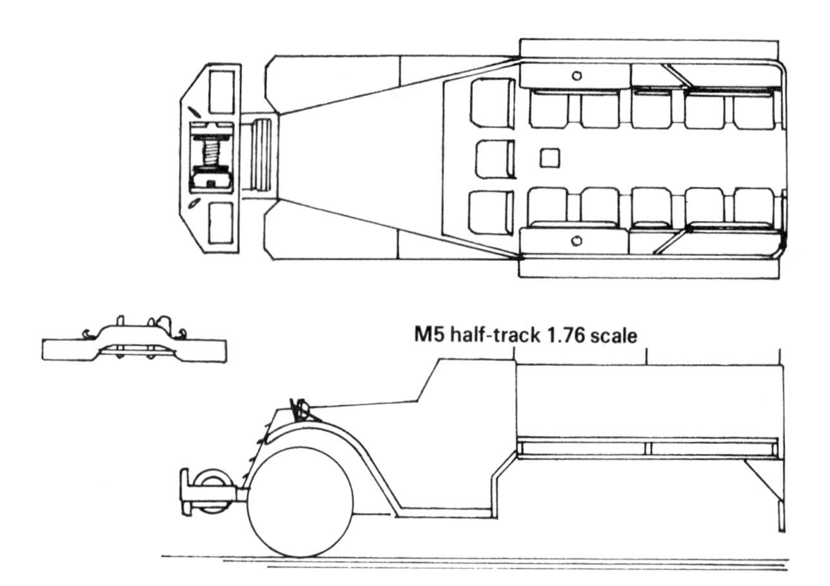

these to the rear of the hull a panel should be fitted behind the side seat backs. The crews' rifles are fitted in brackets between these panels and the hull sides. Seats and seat backs are from 40 thou plastic card with corners and edges rounded off slightly. Note on each side there are two double back rests with a single in the centre. The front seats will require their backs building up and the centre seat needs a complete new back mounted on plastic rod supports. You will also note that on my model I cut away the rear door,

Completed and painted model of M5 half-track.

T12 75 mm GMC
1:76 scale

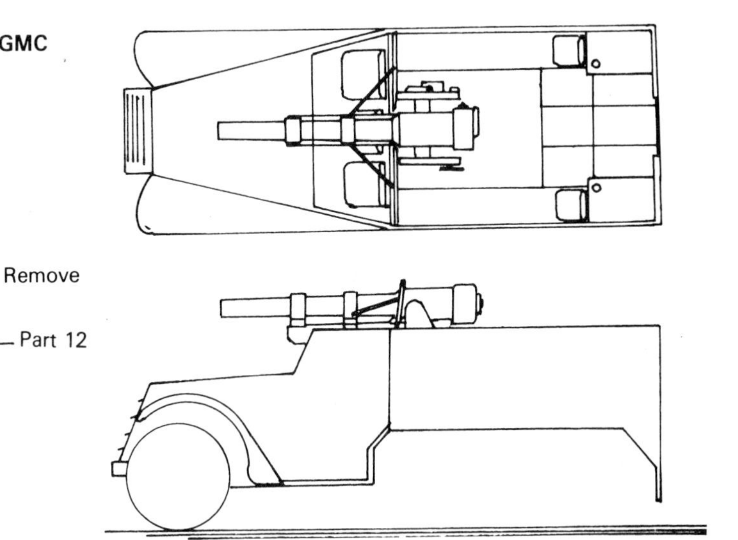

cementing this in the open position. Also a new screen frame was made from Microstrip and the screen also cemented in the open position.

Some of the half-track family were fitted with a power winch at the front instead of the roller included in the kit model. The construction of this is a little complicated but should be clear from the drawings and the sketches. Either laminations of Microstrip or strips cut from thick plastic card can be used for the frame with the drum from scrap sprue etc. Headlight guards were fitted from stretched sprue with the side lights on top of the headlights from short lengths of rod. As these M5s saw service almost wherever there was fighting, colour schemes there are in plenty. The standard colour would be a shade of Olive Drab but there were many applications of additional camouflage colours dependent on the local surroundings.

T12 75 mm Gun Motor Carriage

The half-track family grew, with many conversions to take anti-tank, anti-aircraft, howitzer and mortar armament, and I have tried to model the T12 which carried a 75 mm gun. This was modelled solely from photographs but I have interpreted this into a drawing which can be used as a guide. Unfortunately I was lacking detail of the gun below the level of the sides. With the fighting compartment filled with crew I don't think this lack of authentic detail is too apparent, and if you make the gun readily removable it can always be altered should more information come to light.

The hull modifications entail cutting off the top of the dividing wall (part 12) behind the front seats so that it comes up to the level of the steps in the floor (part 11). The fuel tank (parts 18 and 19) are consigned to the spares box and new ones built up in the rear corners of the hull as shown on the drawing. A new rear floor should be fitted first to fill in to the same level as the steps in the floor (part 11). The seats are fitted facing forwards with their backs to the new fuel tanks. New backs are fitted to the front seats.

The gun uses parts from the 5.5" gun from the Matador kit as shown in the photographs. Shorten the barrel, cut away the portion under the breech (part 61), cut down the mounting (parts 58 and 59) by removing the trail pivots, the balance springs and the outside base so that you can reverse these parts back to front (ie with the barrel pointing away from the end from which the balance

Left *completed model of the T12 75 mm Gun Motor Carriage with Airfix paratroopers standing in as crew figures. Five or six figures are really needed.*

Centre *new fuel tanks in the rear of the T12. Alteration to the driver's bulkhead necessitates new seat backs.*

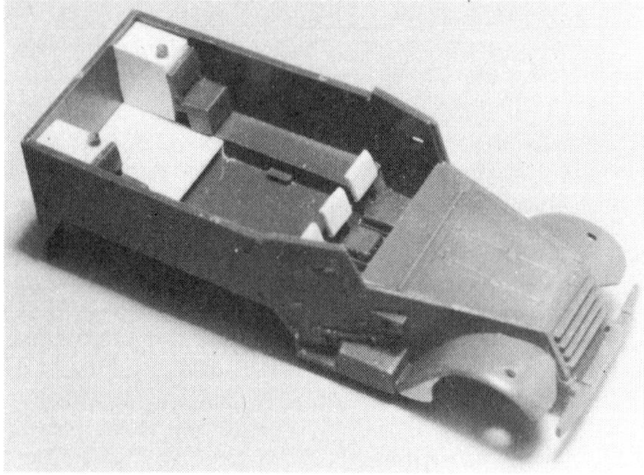

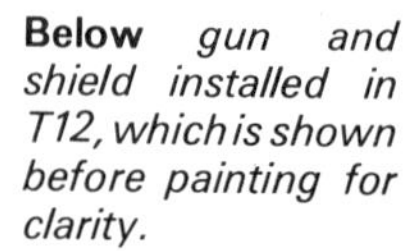

Below *gun and shield installed in T12, which is shown before painting for clarity.*

springs were removed). Perhaps the sketch and photographs will help unravel this better. The mounting I salvaged from an old 88 mm gun (part 91).

I also decided to make this a White or Autocar type by altering the mudguards around the wheel axles, and rounding off the outside edges. They shouldn't strictly be square to the bonnet sides but this can be disguised by stowing the usual roll of camouflage net in this area. Other International variants are the M9A1 which had a different interior layout and the machine-gun 'pulpit' (parts 40-42 in the kit), and the M14 Multiple Gun Motor Carriage (twin .50 cal machine-guns). White type conversions are M4A1 and M21 81 mm Mortar Carrier, T30 75 mm Howitzer, T48 57 mm Gun Motor Carriage, T19 105 mm Howitzer and many workshop, recovery and other specialist types. A profitable field for some research of your own.

88 mm Flak gun and tractor

Like the American half-track family the German Mittlerer Zugraftwagen 8-ton SdKfz 7 was used as a basis for a lot of specialist jobs apart from its designed purpose as a gun tractor.

SdKfz 7/6 Mittlerer Flakmesstruppwagen

One of the specialist bodies fitted to the SdKfz 7 was that to carry the 13 men of the survey team atached to Flak units with a large instrument store compartment fitted at the back. As you will see from the drawings and photographs this is quite a simple little conversion once you have managed to cut out the new sides to the crew compartment. The chassis can be made up as Section 1 of the kit instructions, painting as you go, of course.

The main body floor (part 46) should have the seat location bars removed and parts 44, 45, 47 and 51 consigned to the spares box. The bonnet and cab front can be assembled to the main floor. Cut out the two new side pieces to the dimensions of the drawings using 30 thou plastic card and fit these into position, making sure that they are both vertical. Cement the seats in place (you will have to use one from a second kit or make one up from plastic card), spacing them neatly between the sides. You can leave the rear seat until the storage compartment has been fitted, and this should present no problems as it is just a box shape to the size shown on the drawing. Cut the hood (part 66) to fit and when the interior has been painted, cement in place with new rear side curtains from tissue paper.

Access steps and the front bumper framework are from thin plastic rod with the width indicators on each mudguard from stretched sprue, with the knob formed by holding the end near a candle flame until it melts and rolls back. I hung a spare wheel rim (note the 'spokes' should be removed) on the rear with a camouflage net on the stowage compartment. The trailer is a 'captured' American from the M3 set but this does look very much like the German type often towed by this kind of vehicle.

Mittlerer Zugkraftwagen 8-ton mit 3.7 cm Flak 36

One of many anti-aircraft and anti-tank guns fitted to the SdKfz 7 was the 3.7 cm Flak 36, early versions not having the armoured cab as depicted in the

Flakmesstruppwagen
1:76 scale

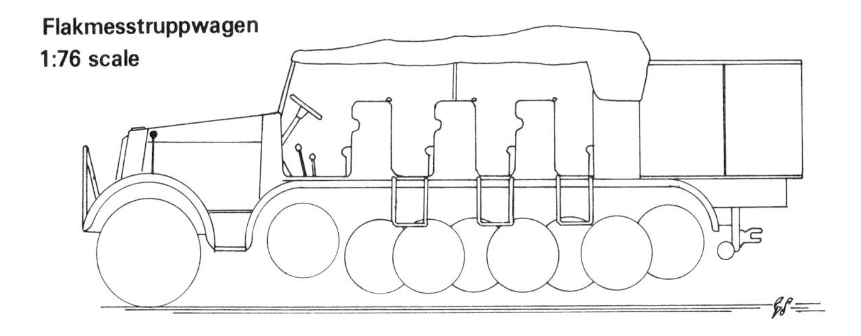

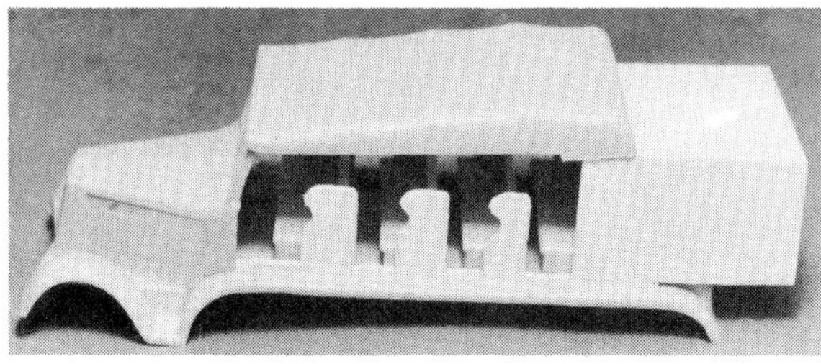

Left *new body sides and extended rear compartment for Flakmesstrupp-wagen.* **Above** *finished model with 'captured' trailer. Note spare wheel is a rim and tyre only - the 'spokes' are removed.*

drawings and the model photographs.

First job is to remove the seat locating bars from the floor (part 46) as we shall fit a new floor over the top up to the back of the front (driver's) bench seat. Fit one of the remaining seats facing the rear with its back up to the back of the drivers seat. Cut away the front of the track mudguards where they come over the sprockets and down to the step. Assemble the bonnet top, front and sides and cement in position to the floor sections (part 46). From plastic card cut a new cab front to replace part 56 and commence building up the cab face by face. You will find it easier if you do the

Zugkraftwagen with 3.7 cm Flak 36, showing crew figures doing a little maintenance.

Zugkraftwagen mit
3.7 cm Flak 36
1:76 scale

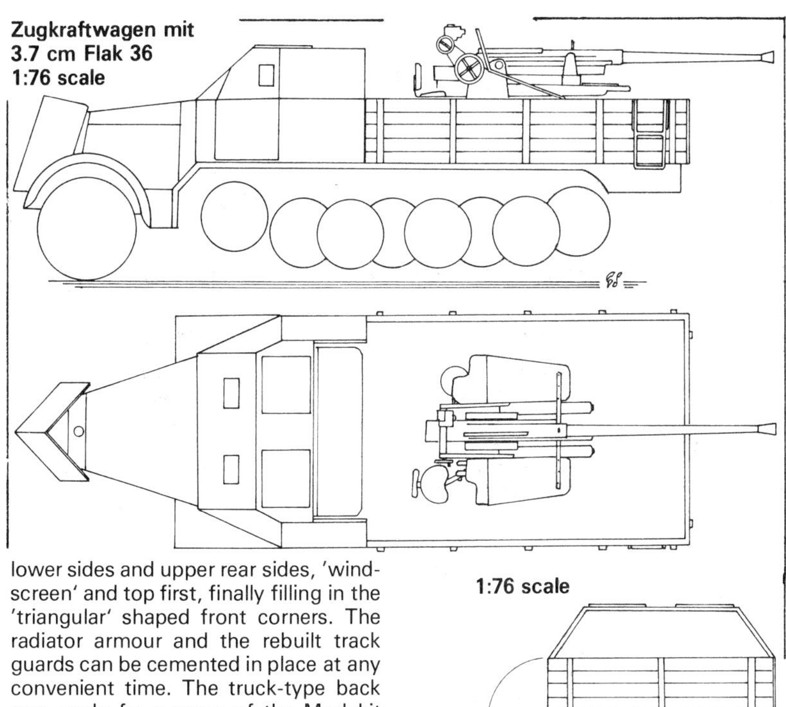

1:76 scale

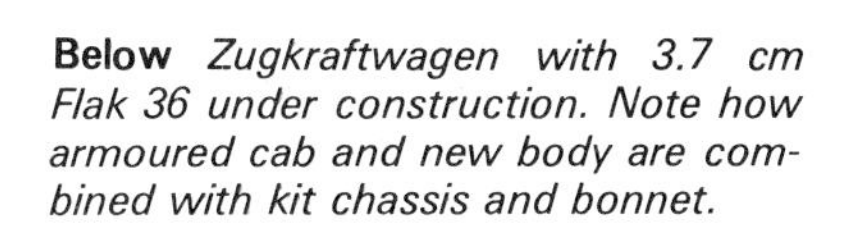

lower sides and upper rear sides, 'windscreen' and top first, finally filling in the 'triangular' shaped front corners. The radiator armour and the rebuilt track guards can be cemented in place at any convenient time. The truck-type back was made from some of the Modakit planked plastic card with metalwork from Microstrip. Note the extension to the rear 'mudguard'. The 3.7 mm gun came basically from the SdKfz 234 Armoured Car kit using the barrel assembly, with the end reshaped, and the mountings (parts 1,2 and 4). The rest was built up from scraps of plastic card with handwheels from the '88' etc. The seat and armour shield are from plastic

Below *Zugkraftwagen with 3.7 cm Flak 36 under construction. Note how armoured cab and new body are combined with kit chassis and bonnet.*

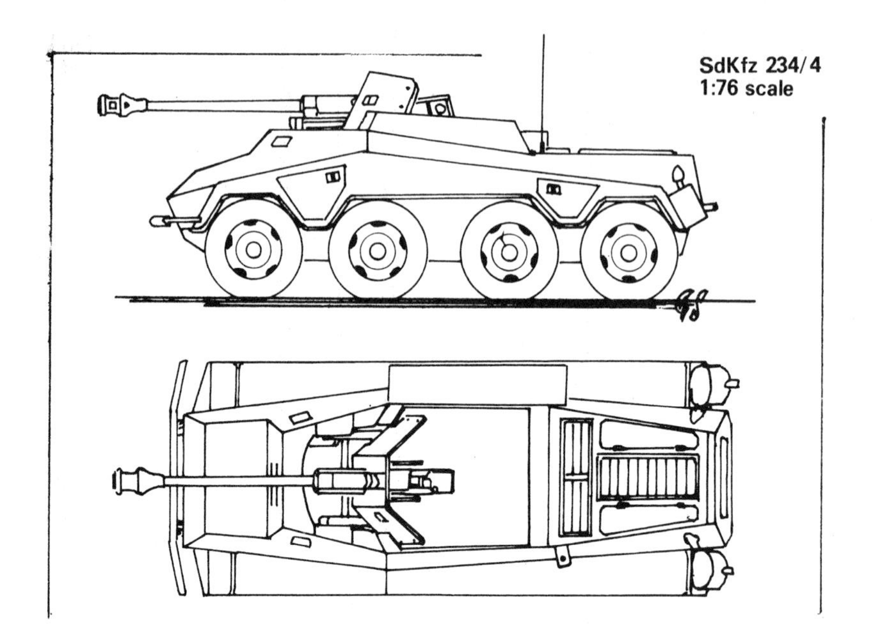

card. It is often difficult to remember where the odd little parts do come from but as you build up your own spares box so these bits and pieces will become available. There does not seem to be a lot of detailed information available on some of the guns of the Second World War but in this small 1:76 scale it is not difficult to obtain a reasonable likeness just by reference to photographs.

The photograph of the finished model is painted in the usual sand (Humbrol HM 2) with a reddish-brown mottle camouflage. The crew are doing a bit of maintenance but if you require an in-action crew those from the SdKfz 234 kits could be utilized.

SdKfz 234/2 and 234/4

The Airfix kit of the 234 'Armoured Car' is unfortunately a bit of a mixture of early and late models. The main error is in the mudguards as the basic shape of the hull is quite acceptable. For all 234 models therefore we must make new mudguards as shown on the drawings. To simplify this I have drawn these parts actual size, but remember you may have to bevel adjoining edges to get a perfect fit. The tops were cut from 30 thou plastic card with the sides from 10 thou, using scrap strips underneath to make a stronger join. Note the rounded edges to the completed mudguards. If you decide to do the full 234/2 Puma conversion then you want the four lockers scoring in the side—the 234/4 has the two centre ones blank.

The top of the engine deck (part 14) should have an additional transverse grille and this I fitted (from Modakit planked card) after carving away the moulded hinges. The exhaust pipes (parts 20 and 21) as in the kit can be fitted to lead into large cylindrical silencer boxes mounted on the rear of the mudguard.

If you are making the 234/4 version then that is all the modification required although you may like to add a tubular 'bumper' across the front and a spare wheel at the rear. A stowage box as shown on the drawing was fitted to the offside.

The 234/2 'Puma' entails rather more work, of course, starting with the removal of the side armour that sticks up each side the rear half of the fighting compartment on parts 10 and 11. I fitted a new top, rear and extension to the engine deck before filling in the triangular rear sides but this should be visible

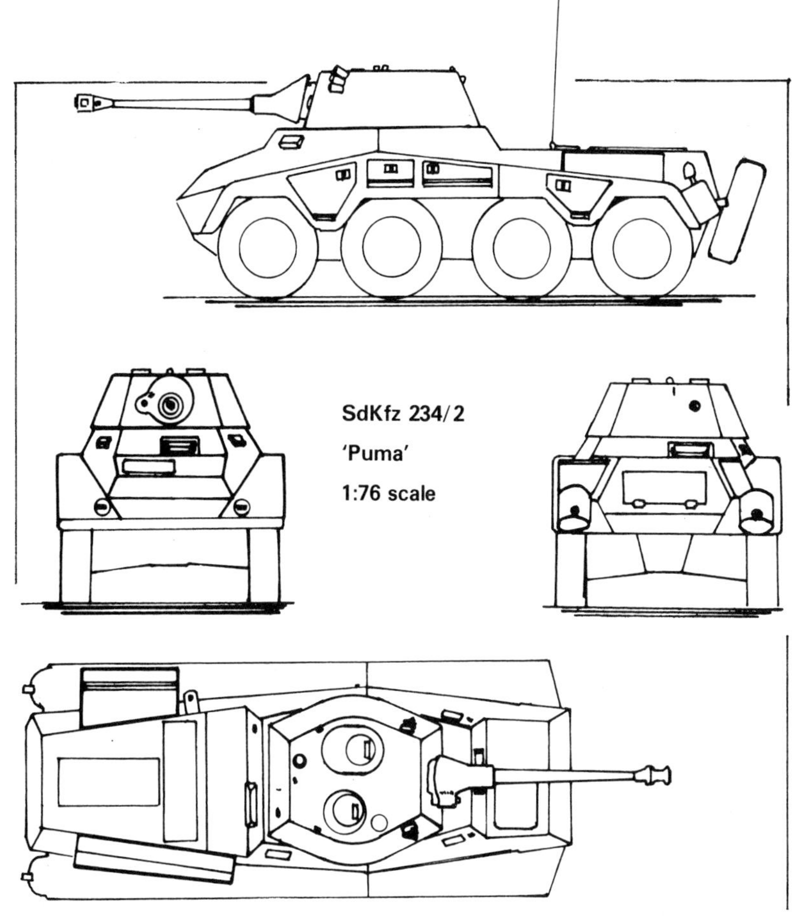

on the photographs of the model in construction. Note also the extra stowage boxes on the mudguard top. The turret is constructed as the photographs using 20 thou plastic card for top, bottom, front, rear and the internal strengthening partitions. The sides were then cut roughly to shape (from 10 thou

Left-hand side view of the model under construction clearly shows the extensive modifications which are necessary to produce a decent Puma from the Airfix 234 armoured car kit.

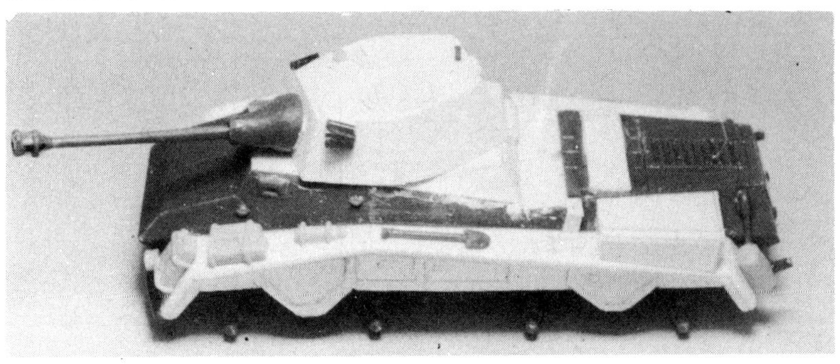

plastic card) and curved round this framework holding in place with strips of Sellotape. Mek-Pak was then brushed into the joins and when dry surplus plastic cut away. Hatch covers are from 10 thou card with smoke discharges from plastic rod. The gun is from the Stug III kit as is the mantlet which requires a lot of re-shaping to get it down to the Puma's size.

That concludes this short chapter on some of the half-track and armoured car conversion possibilities from the Airfix range of 1:76 scale kits. This does leave a shortage of Allied armoured cars but these have to be completely scratch-built using just wheels and other odds and ends from Airfix kits. I have already covered some of these, like the Staghound and the Marmom Herrington, in articles in *Airfix Magazine,* but they are really outside the scope of this book which is meant for the beginner to kit converting. However, I would suggest that as soon as you feel confident, and have had experience doing some of the conversion work I have been describing, then send for some of John Church or Len Morgan's plans and make the attempt.

Top *view of right-hand side of model under construction.* **Above** *method of making the turret.* **Below** *completed Puma model.*

four

Tanks and other fully-tracked vehicles

To work through all the possible conversions from the Airfix range of tank kits would need a far bigger book than this and I have therefore had to select just a few as a sample. I have included some 'quickies' which only take very little longer to do than the basic kit, others will take you a lot longer and I hope test your skills a bit.

Probably the most used and abused tank of the Second World War was the Sherman, and though the Airfix kit has been in the shops for many years now, it maintains its popularity with conversion addicts because of its many possibilities.

From its introduction it was improved, up-gunned and up-armoured, formed the basis of specials like mine clearers, bulldozers, artillery tractors, armoured recovery vehicles, self-propelled guns and inumerable 'odd ball' experimentals. Some of these are extremely complicated conversions and outside the scope of this present book so I will start this chapter with some easy ones.

Reworked M4

The need for a bit more protection soon became apparent as the opposing guns got more powerful and this was added quite simply by welding plates of 'applique' armour to the more vital areas. The drawing shows the shape of these and the photographs of my model in construction shows them fitted to the model. All are from 20 thou plastic card which you will find can be bent to the correct curve for the turret and new gun mantlet without any trouble. Sellotape

The reworked M4 model.

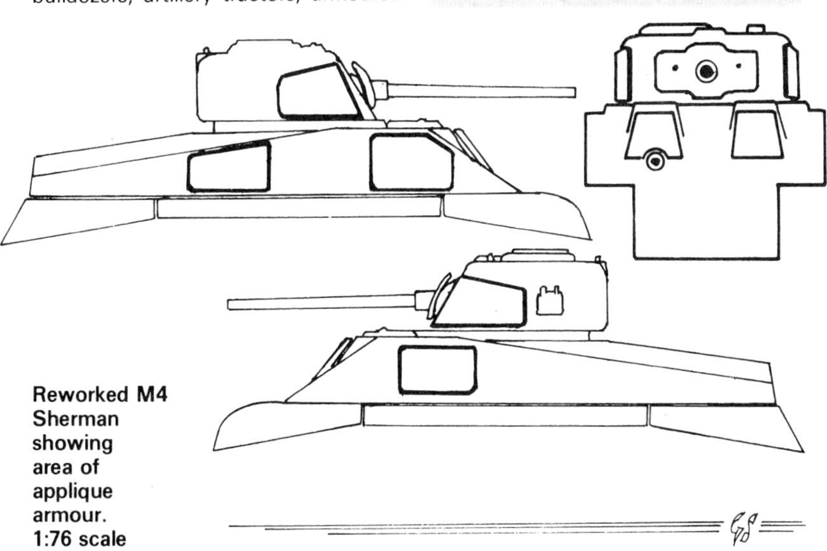

Reworked M4 Sherman showing area of applique armour. 1:76 scale

Top *the white plastic card parts show where the applique armour is added to the front, turret and port side of the reworked Sherman.* **Left** *the starboard side of the tank.* **Below** *completed, painted model. The addition of dust shields would make the M4 even more different.*

will hold them in place while the adhesive dries. Some vehicles were also fitted with full sandshields or with the small triangular front 'mudguards'. These again are options you can incorporate on your particular model if you wish.

M4A2 British Sherman III

The M4A2 was fitted with a General Motors diesel engine in place of the Continental 9-cylinder radial of the M4 and this meant a revised rear engine deck layout. The first job therefore is to carefully remove the unwanted detail moulded on the kit and fill any unwanted score marks etc with body filler and then add the new grille either from scored plastic card or some of the Modakit planked card.

Hinges are from Microstrip and filler caps punched from plastic card. My particular model is fitted with the early type suspension which features return rollers on top of the suspension brackets and of course these came from a Lee/Grant kit. Fill in the Sherman suspension locating holes in the sides (parts 4 and 24), cut off the Grant suspension locating pins and cement the complete assemblies in place direct to the side pieces. The Sherman suspension can similarly be fitted to the Lee/Grant hull to give a later model of this tank so nothing is wasted here.

The turret can be made up and fitted with a stowage bin on the rear (I used a spare from a Crusader) and the model assembled and painted up to the stage

Above *view of the completed M4A2 (British Sherman III) fitted with sand shields, side and camouflage netting etc. Note Crusader stowage box on turret and the Lee/Grant type suspension.* **Right** *close-up of the new rear engine decking on the M4A2 model before painting.*

shown in the photograph. The tracks should also be pained and fitted to the model when both are dry–meanwhile you can be cutting out the sandshields from 10 thou plastic card. The side netting and rolled up netting on the rear deck were made from some old nylon. The side nets are from narrow strips tied at intervals with cotton. The sandshields can then be fitted in place on the model and painted before the final details like the side netting, crew and other bits and pieces are added, finally touching up with paint if required.

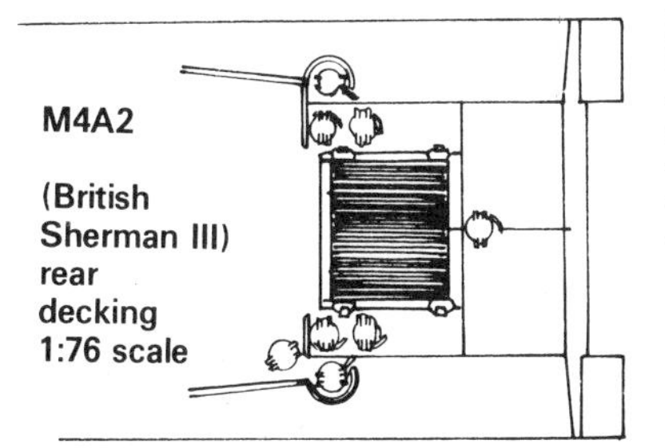

Crusader

This has long been one of my favourite tanks and although the Airfix kit is a little too long it goes together well and looks like a Crusader. Probably this tank is most famous for its role in the Desert battles with Rommel, for which its speed was admirably suited if not the power of its armament. The Desert battles were also a battle of wits, not just a slogging match, and one of the devices designed to deceive the enemy was to fit a tilt cover to the Crusader so that it looked like a lorry. This disguise was aided by blacking out the three centre roadwheels.

To re-create this on the model, first assemble the Crusader II as the kit instructions and then build up the frame to the dimensions shown on the draw-

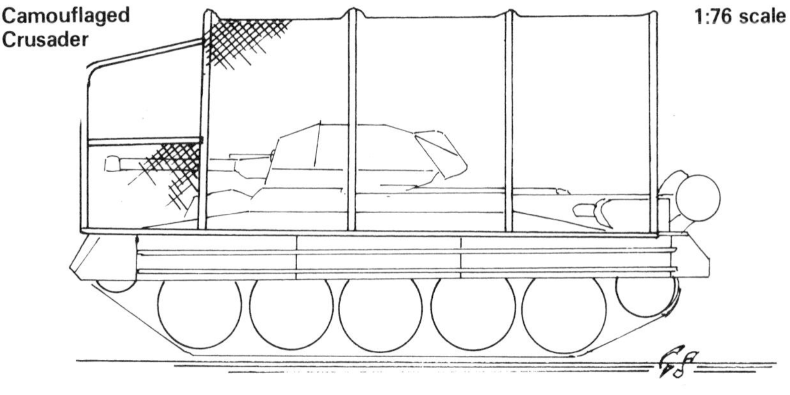

ing. You will find that the plastic rod bends easily round a pencil or paint-brush handle, the only difficulty really is getting four the same. Cement the longitudinal rod along the sandshields first and when dry erect the hoops, then the top rails. When these are all dry the completed framework should be painted and then the tilt netting cemented in place. This was often quite torn and tatty so it doesn't have to be too neat to look authentic. When dry this can be painted and any other touching up done at the same time.

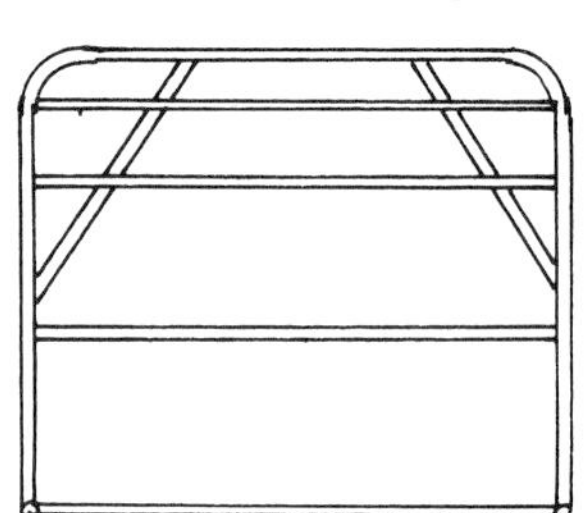

Crusader anti-aircraft tanks

The invasion of Europe anticipated trouble from the opposing air force on D-Day and for some time after, before airfields could be set up as bases for the Allied fighter aircraft, and to combat this problem several designs of tanks fitted with anti-aircraft guns were made ready. Among these were two types of Crusader, the Mk I fitted with a 40 mm Bofors gun in a rectangular open turret, and the Mk II fitted with two 20 mm Oerlikon cannon and .303 Vickers machine-guns in an enclosed turret. The hulls in both cases are identical.

Basic framework completed for the disguised Crusader. Note how centre three roadwheels have been painted dark so as to reinforce the impression that the vehicle is a lorry - from a distance.

Completed Crusader model with netting in place - certainly an unusual addition to any model collection and also useful for wargamers, especially if the framework is made removable.

To start with the hull modifications first, carve away the sandshield locating ridge under the hull top (part 20) below the stowage box area. This will make a neater job of this area as only the front and rear will be hidden by the new dust shields. Incidentally, a normal Crusader III can be made without the full-length sandshields. For the AA versions the hatch (part 67) is discarded and rectangular lids from 10 thou plastic card, and hinges from Microstrip, cemented over the aperture.

The Mark I Bofors gun turret is the simplest to make and is a nice easy lesson in marking out for shapes with sloping sides. You will realise that it is not possible to trace the shapes of the sides direct from the drawing otherwise you will finish up with a turret that is too squat, and in any case they wouldn't fit together. The side panels should be drawn onto the plastic card, taking all measurements for length from the side view but measurements for height from the front and rear views. I have left in some construction lines as a help with this. The front and rear are similarly drawn on the card but taking height measurements from the side views. Before cutting out the parts, check that the lengths of the front corner edges agree on both side and front plates. Similarly mark out the rear plate but allow for the thickness of the plastic used at the sides as this rear plate fits

Crusader AA tank with new Bofors gun and modified side shields to the running gear.

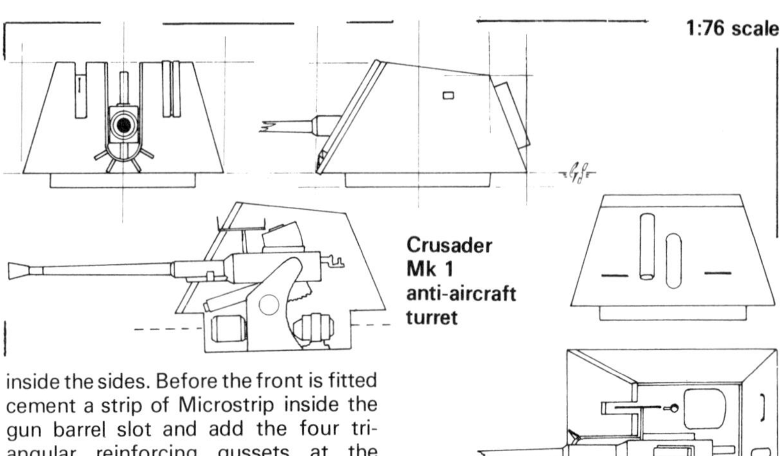
1:76 scale

Crusader Mk 1 anti-aircraft turret

inside the sides. Before the front is fitted cement a strip of Microstrip inside the gun barrel slot and add the four triangular reinforcing gussets at the bottom.

The actual gun is again a mixture of bits from the spares box, Plastruct tube and plastic card parts, but the drawing will serve as a guide.

The front and rear dust shields are from 10 thou plastic card to the dimensions and shape shown on the drawings.

The Mark II Oerlikon turret was first published in *Airfix Magazine* in April 1972 and the drawings and sketches have been repeated here for the benefit of those who haven't got the original. To summarise the instructions, first cut out a base (30 thou plastic card) 15 thou undersize all round except at the front where it is 30 thou undersize. This is to allow for the thickness of the sides and front, a point which should always be

Above *construction of Mk 1 Bofors turret. White areas are plastic card, dark sections are pieces of kit scrap.* **Left** *Crusader model with Mk 2 AA turret constructed as shown in the drawings on the facing page.*

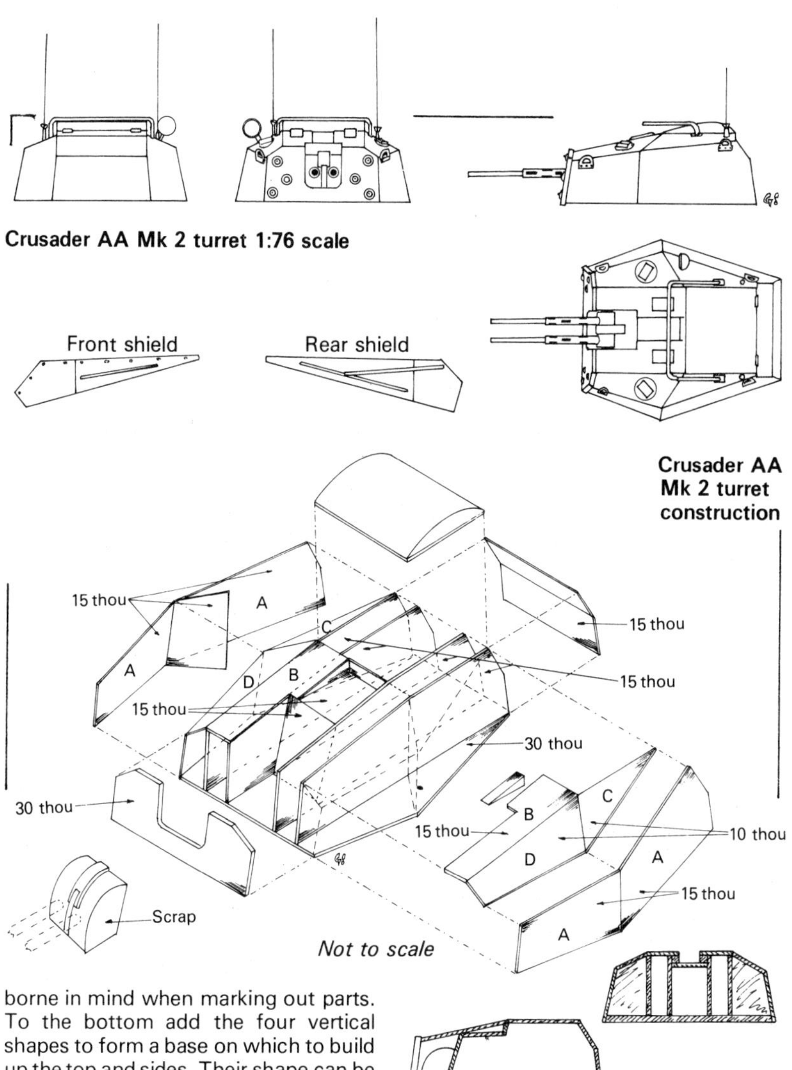

Crusader AA Mk 2 turret 1:76 scale

Crusader AA Mk 2 turret construction

Not to scale

Cross section of turret: thickness of sides not to scale

borne in mind when marking out parts. To the bottom add the four vertical shapes to form a base on which to build up the top and sides. Their shape can be taken from the side view but again allow for the thickness of the other parts. Cement them in place to the bottom adding the sides 'A' followed by the top parts 'B' and then 'C' and 'D', chamfering the edges to get a neat fit where necessary. 'C' and 'D' can be fitted oversize and the surplus trimmed off when dry. This is just a brief summary but study the sketch, drawing and photographs for the full construction method.

Churchill Mk I

From North Africa to Berlin there always seemed to be a job for some variant of the Churchill, be it gun tank or specialist vehicle. The many variations and permutations make it a modellers' favourite and the Airfix kit can form a basis for this much used vehicle. As a lot of the specialist uses were built on the early versions I thought that the Mk I should be the subject for this next section, then if you wish to pursue the theme further you will have the necessary knowledge. Basically the Churchill changed little over the years but to take it back to the Mk I does involve a lot of

Left *modified Churchill side plates are shown here flanked by the original components for comparison.* **Above** *and* **below** *two views of the completed but unpainted model clearly showing the extensive modifications needed.* **Opposite** *completed model.*

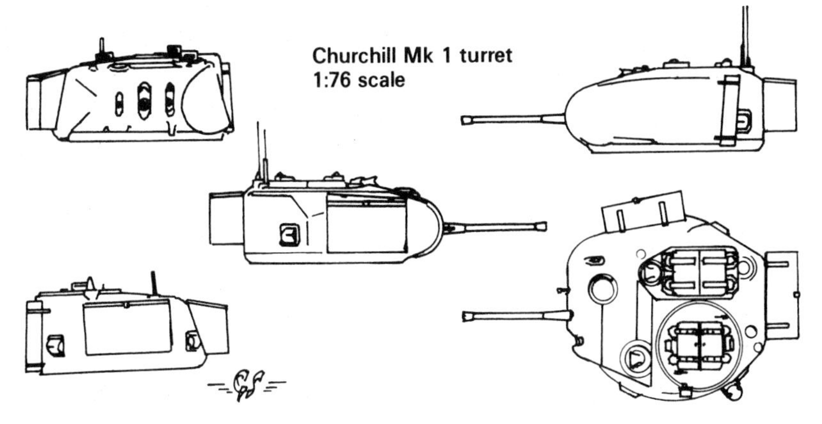

modification of minor points. The photographs should help to show the differences and if you purchase Bellona Print Series 19 the excellent plan by A.J. Gooch will furnish additional detail. I have just drawn the turret as I think the rest can be seen from the photographs.

First cut away the track guards (from parts 42 and 84) and carve away the circular side escape hatch. Cut away the guards from around the front idler. Saw away the top of the inner sides (parts 12 and 54) so that those will be flush with the hull top (part 85) and cut away at the front to match the outer sides. Compare the modified parts in the photograph. You can now complete the assembly as in Section 1 of the kit instructions and as shown on the photograph of the completed hull.

Fit new hatches in place of the circular types moulded on part 87 and note the 'stepped' front with the 3" bow howitzer. Length of the howitzer is 15 mm. The curved top engine air intakes were made from those in the kit (parts 101 and 102) by making the top from sections cut from cylinders from the Emergency Set and fairing these in with body putty. Note the different exhaust system which can be made partly from that in the kit with new pipes from rod.

The turret can either be modified from that in the kit or, if you wish, carved from wood with the detail added from plastic card. I found that a pretty fair Mk I turret could be made by sawing away the rear stowage box, the offside rear face and the nearside front face, filling inside with scrap sheet and when dry, building up the shape with body putty. Filing, carving, sanding and patience do the rest. Detail is added as usual from plastic card parts.

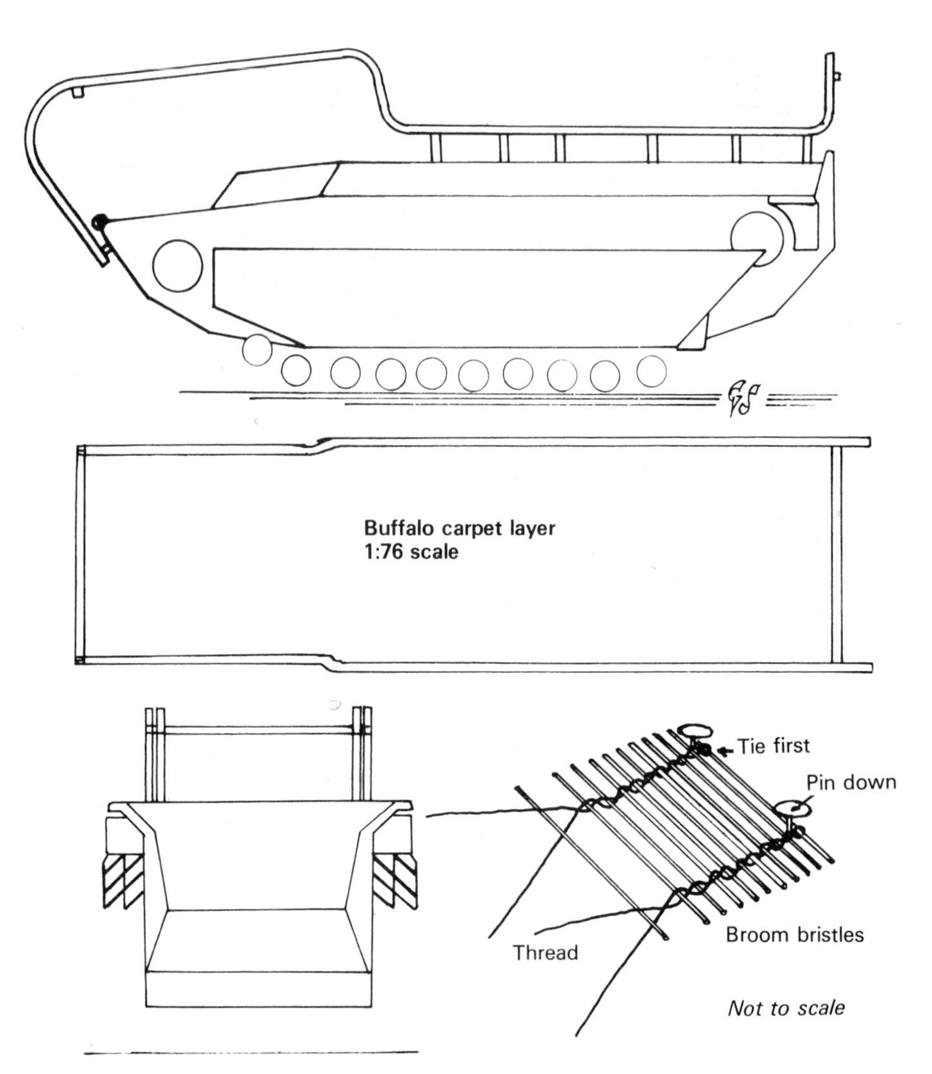

As these early hulls were used for many types of specialist job like bridge-layers and demolition tanks etc, this hull modification will be useful for other Churchill models.

Buffalo carpet-layer

As a result of previous experience the Military found a need for an amphibious vehicle that could lay a 'road' of chestnut paling up river banks to enable heavier amphibians like the Duplex Drive Shermans to get sufficient grip to heave themselves on to dry land. The Buffalo provided the answer as it had good swimming charactersistics and was quite at home scaling muddy and slippery river banks.

The frame to carry the stowed carpet of palings is quite simple to construct from lengths of Microstrip and the photograph will show how I did this by pinning the lengths down over grease-proof paper placed over a tracing of the plan. Two laminations were used, Mek-Pak was brushed between and the two laminations allowed to dry. You will find that when removed from the board they

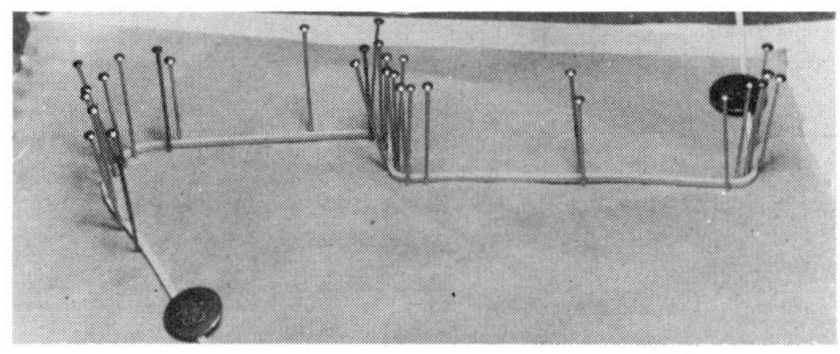

Pinning out the Buffalo carpet support rails from plastic rod.

Carpet supports fitted to Buffalo hull.

Completed model with carpet of broom bristles in stowed position.

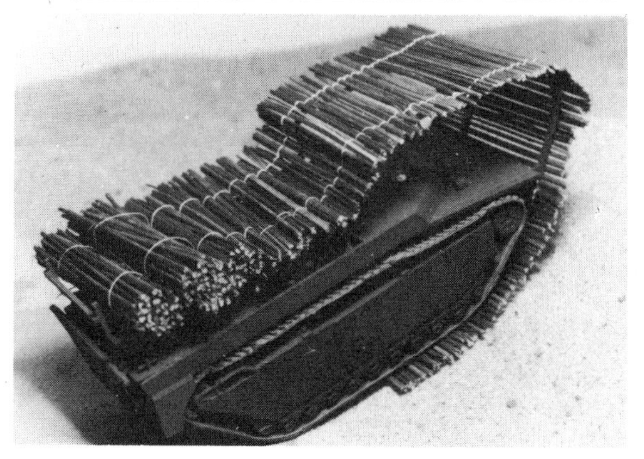

Starting to lay the carpet. Note the rear vanes.

will retain their shape. These are then mounted on the supporting strips cemented inside the hull sides. Construction of the hull is as the kit instructions but leaving off any armament. A small detail improvement is the fitting of rear vanes each side the hull just behind the drive sprocket. Overall colour of the model is Olive Drab.

The chestnut paling carpet is probably one of the most tedious modelling jobs I have tackled for a long time but this was eased somewhat by the help from my son, Nicholas, who placed the broom bristles that I used in place while I manipulated the threads. The sketch shows how this was done by weaving the thread between palings, tying off after about each four until a length sufficient to drape over the bow and back to the stowage had been accomplished and thereafter tying them in bigger bunches to rest in the stowage area. An alternative to this is to cut a strip from a rush or bamboo table mat if you can find one with suitable thin slats.

Matilda bulldozer

Across the other side of the world from the European battles, the Australians were among the Allied troops campaigning against the Japanese Army in the jungle. Matilda tanks were among the few types which were operated in this area when the need arose for a bulldozer to clear a path through the jungle. Early modifications to fit a blade on the Matilda were a bit primitive with winch operation of the blade arms. The developed version had hydraulic ram operation and was equipped so that it could jettison the arms and blade, back out and immediately become a fighting tank with all the capabilities of a normal Matilda.

If you want to try a complicated conversion then have a go at making a working version as I did--it is a test of patience! The alternative is of course to cement the parts in place with the blade and arms in either the up or down position.

Follow the kit instructions to make up the complete hull at least to Stage 3 but carve off the spare track links from the front mudguards (parts 41 and 42). To avoid accidental damage leave off the smaller detail parts. The turret is standard as the kit except that the commander's cupola (parts 56-57) is reduced in height as shown in the photographs.

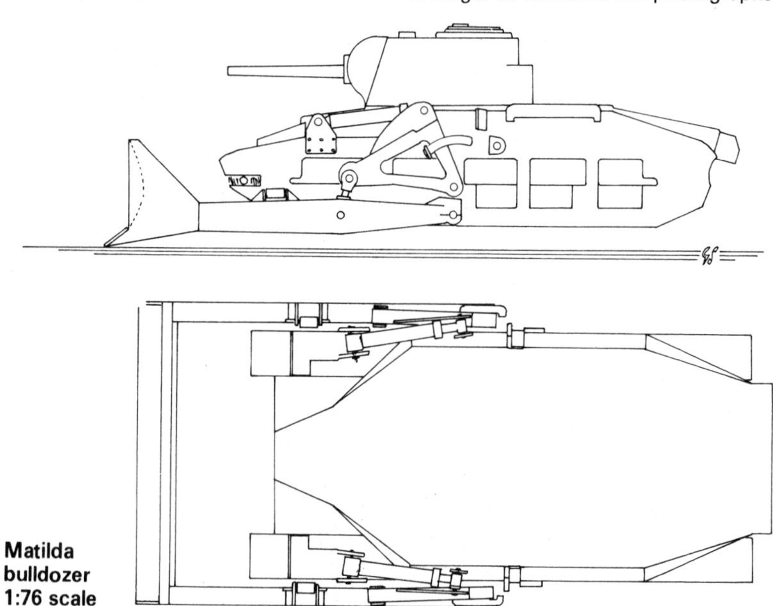

Matilda bulldozer 1:76 scale

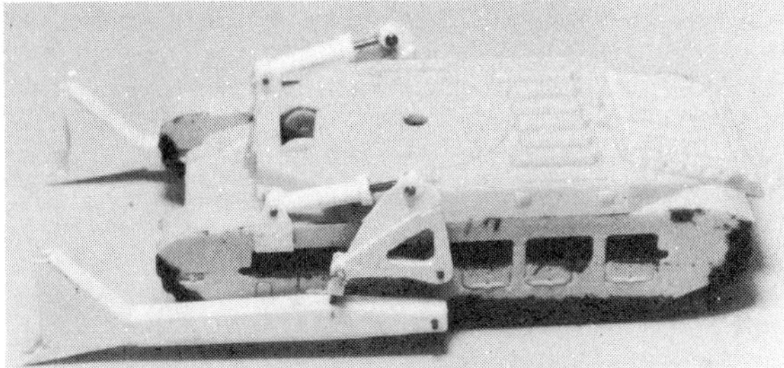

Right *Matilda dozer rams and arms fitted in the down position.* **Above** *blade has now been added together with other detail. The blade is raised and linkage up to the limit stops on hull sides.*

Completed model on small base: I hope to add an Australian crew and support infantry to this model to complete the scene.

Follow the usual routine of studying the drawings, sketches and photographs before commencing construction.

I split the assembly up into three sub-assemblies for each side consisting of rams, arms and the intermediate link fabrications. The rams are from Plastruct tube with pistons of plastic rod. The arms are built-up boxes which are made by cutting out four pieces to the side view shape, making certain that they are all identical. Drill the end pivot bearing holes. Cut away to receive the curve of the blade on the two inner pieces. Cut a long strip of plastic card to use for the top and bottom of the box section and cement these together, keeping square. The intermediate link assemblies are from two laminations of 30 thou plastic card as this is easier to work than one of 60 thou. Drill for all pivot bearing holes. The rods on which all the moving parts pivot have a 'head' formed on the end by holding it up to the side of a candle until the plastic melts and rolls back.

Keep the rod turning to obtain a true head. Construct the brackets to hold the ram cylinder ends to the top of the track guards/hull.

When all six sub-assemblies and the pivots are ready, drill mounting holes in the hull sides as required and try a trial run to see that everything fits and will operate satisfactorily, adjusting if necessary. When satisfied that everything is correct cement the guide rollers to the arms and the limit stops to the hull. You may prefer to paint all parts separately. This is the best idea as they tend to gum up if painted *in situ* and in this case remove them and paint before re-assembly. The blade portion is fitted after the arms, rams, etc, are all in place, starting with the curved main blade from 10 thou plastic card with the top from Microstrip and the bottom edge from a strip cut from 20 thou sheet. When dry the rest of the detail parts are added and the final painting, crew and weathering finished off as required.

Panther Ausf A,D and G

The Airfix Panther kit was among the early releases in the Armoured Vehicles series and unfortunately contains a few errors. Some of the best references available on the Panther models are contained in the Bellona Series of which Number 16 includes the Ausf G and Number 26 the Ausf A and D. The Airfix kit most nearly resembles the 'G' model and to make this more accurate refer to the details drawn in on my outline drawing. These are the various vision blocks on top of the hull and the escape hatches, the deletion of the vision aperture on the nose (next to the bow machine-gun ball mounting is removed, stowage boxes and the gun. Other modifications which affect all models are the angle of the hull front, the 'sit' of the model, and the turret, so these will be dealt with first.

The Airfix model unfortunately sits too low but, depending on which type you intend to model, there are two fairly easy ways of correcting this. Both are illustrated by sketches and photographs. You will also have seen that

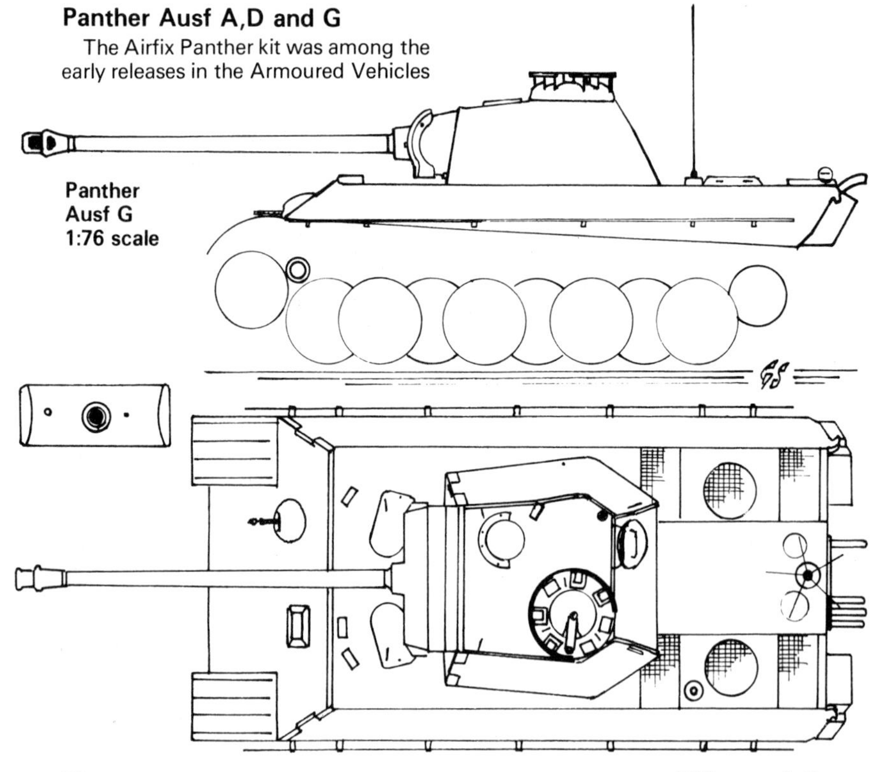

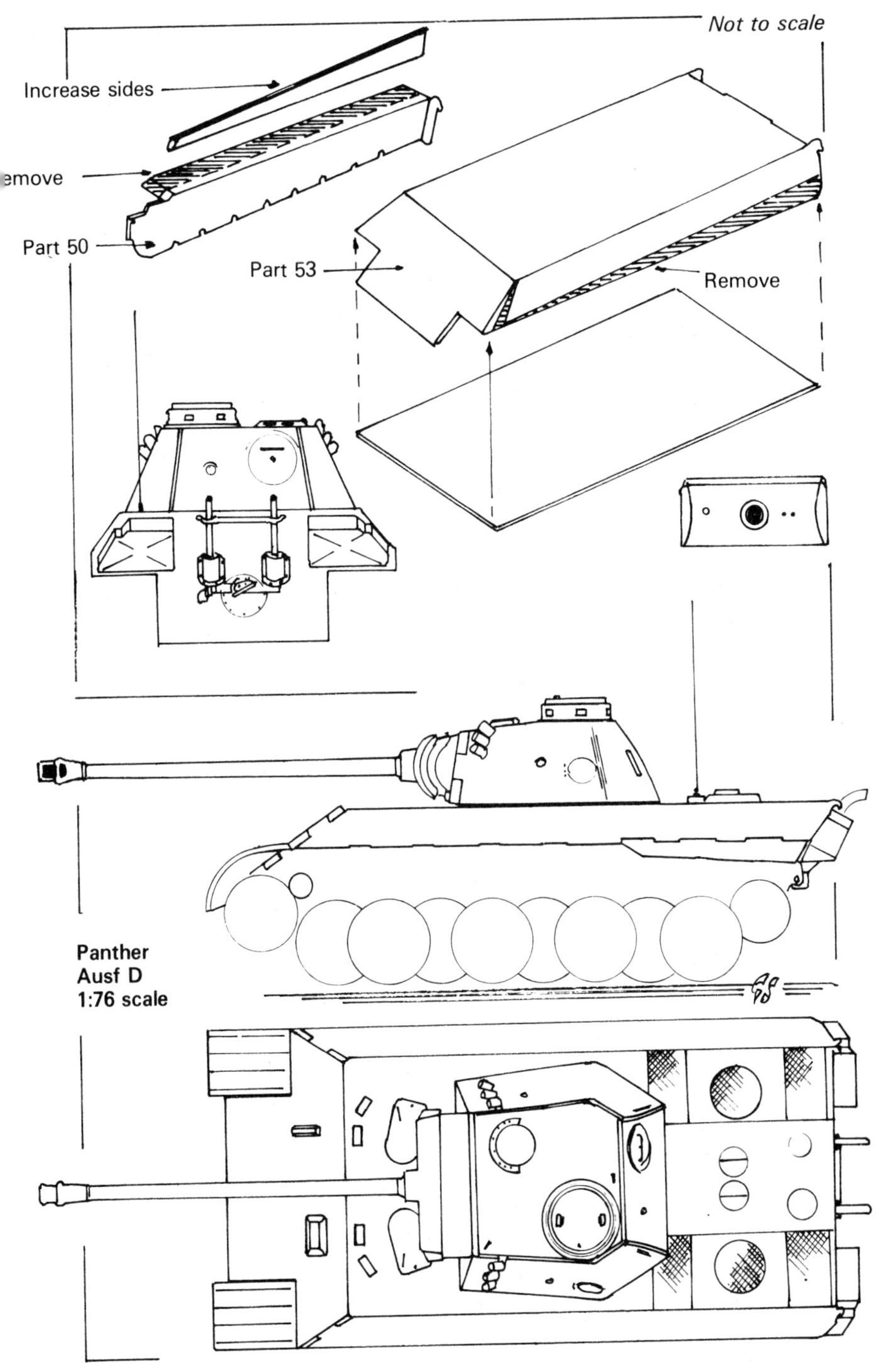
Not to scale
Increase sides
emove
Part 50
Part 53
Remove
Panther
Ausf D
1:76 scale

Panther Ausf A
1:76 scale

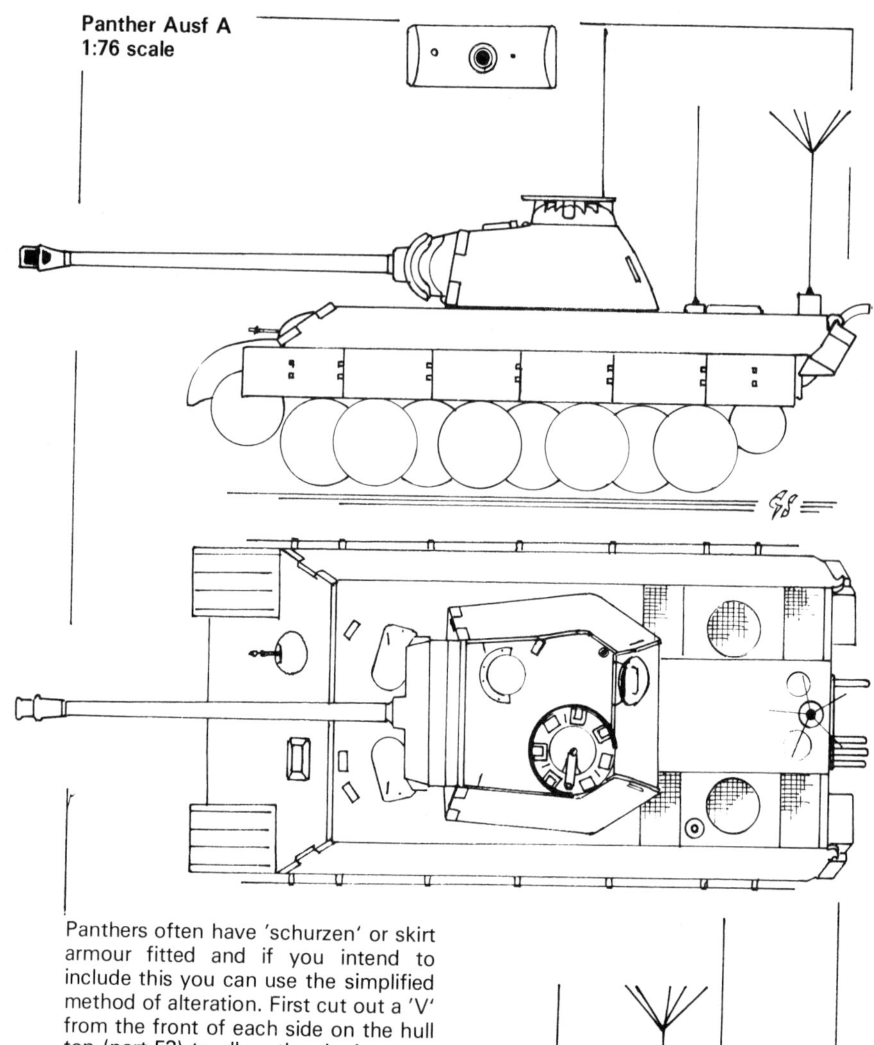

Panthers often have 'schurzen' or skirt armour fitted and if you intend to include this you can use the simplified method of alteration. First cut out a 'V' from the front of each side on the hull top (part 53) to allow the sloping nose plate to be bent to its correct angle. Next cut a sheet of 20 thou plastic card to fit just flush inside the underside of the hull top. Cut off the top flange of the two hull sides (parts 50 and 51) and add the additional sections cut from 40 thou plastic card to bring these sides up to the correct height. If you are a stickler for detail you will note that the rear idler is just a little too low (about 1.5 mm) so you can remove the boss and relocate this up the hull side. The parts can then be fitted together as the kit instructions,

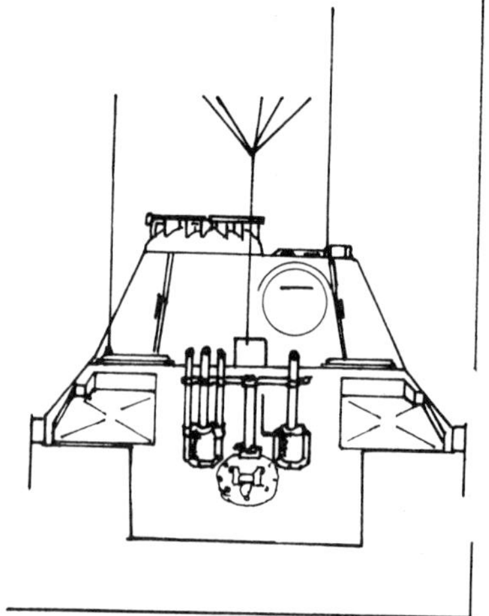

This photo clearly shows the alterations to the Panther hull parts required to make the Ausf D and A.

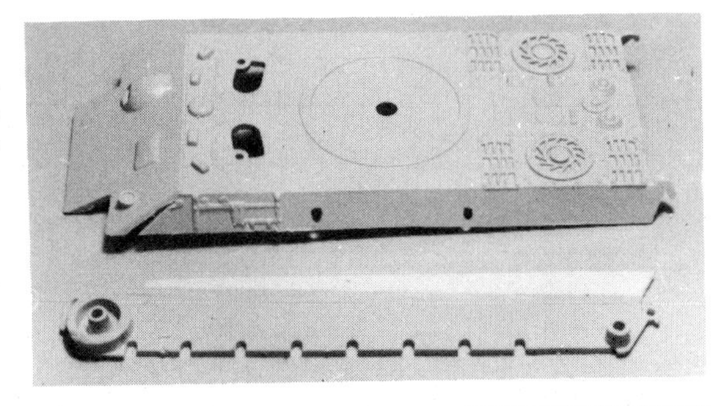

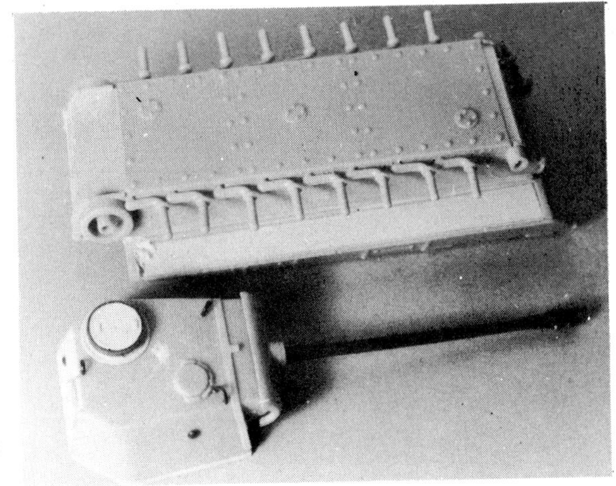

This shot shows modifications to the Panther turret and hull.

filling any gaps with body putty and inserting an extra section at the rear where part 54 should join to part 9.

The Ausf A and D models you will see have different sides to the hull top, these being parallel with the top deck so this should be cut down before the underside of the hull top is sheeted over.

The general size of the kit turret is slightly too small and to adjust this we add a sheet of 30 thou plastic card to the bottom and a sheet of 15 thou to the rear face. A mounting ring for the machine-gun can be added to the cupola (part 76).

I think the kit gun barrel also looks too thin for a Panther and replaced mine with a reworked Stalin III barrel. Note the different shape at the bottom of the Ausf G mantlet.

Ausf D

I decided to turn my Airfix Panther into the Ausf D with the addition of skirt armour and thus used the first method to alter the sit of the hull. The hull machine gun ball mounting is removed the kit driver's entrance hatches and the exhaust system retained. A new turret cupola was made from part of a spare Sheridan road wheel and discs from plastic card, and was substituted for the kit cupola. Other turret detailing was as the drawing. Rear stowage boxes are built up from plastic card and front track guards and skirt armour fitted. The various engine grilles should strictly have mesh covers but I have not yet found any that is fine enough.

Ausf A

Briefly, the differences between Ausf D and A are in the turret cupola, which is as the kit (part 76); and the exhaust system, which has the three pipes on the nearside as in the Ausf G. Note the detail differences on hull top and turret. It should also have wire mesh grille

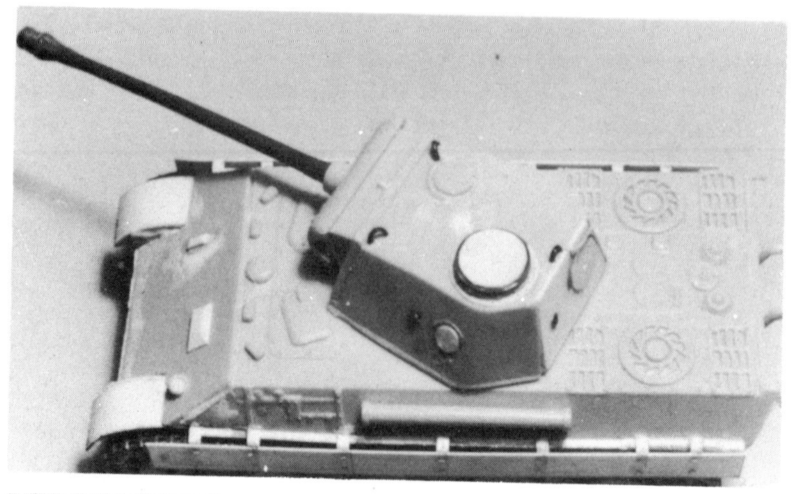

Above *skirt armour and mud guards added to the Panther Ausf D model.* **Left** *the finished model in a plain sand colour scheme with much muddied lower hull.*

covers. Skirt armour and/or front track covers are optional as these were often wholly or partly lost in service.

Panzer IV

The Airfix PzKpfw IV must be the most convertible of all the Airfix range as it was used as the basis for many self-propelled guns, anti-aircraft guns and other specialist vehicles. Many of the parts can also be used on PzKpfw III conversions, of course, and to avoid waste it is a good idea to plan your conversions to use up left-over bits. For example, turret hatches, cupola, gun parts, etc, of the IV can be used on the III while the rest of the hull is used as a base for a self-propelled gun. A good example of this is the Sturmgeschutz IV (L/48) which utilizes StuG III top direct on the PzKpfw IV hull.

Sturmgeschutz IV (L/48)

Two sketches copied from the kit instructions show the parts that are used for this simple conversion project. First then, cut away part of the PzKpfw IV hull top (part 78) and assemble the complete hull, including tracks, painting as you go. To ensure the StuG III hull fits snug on the top you will have to remove the PzKpfw IV hull top side locating bars at the front only. After a trial fit cement the StuG hull top in place and fill the gap between its front and PzKpfw IV front with a strip of plastic card. I used part of the Panther gun (which you may remember was left over from the Panther modifications where I had used a JS III gun) as I thought the StuG gun far too thin. Build up the extended driver's compartment from plastic card with hatch etc, added from scrap. If you

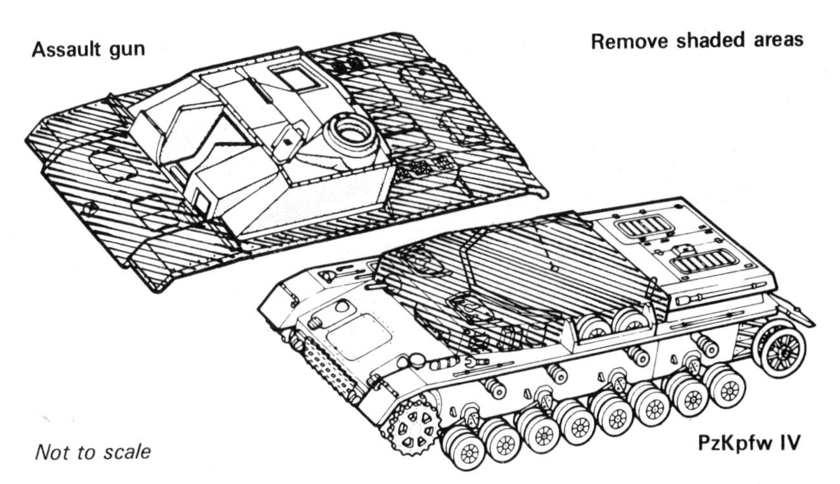

wish you can add either the side-hung armour attachment frame or the complete armour and any other detail you wish—like additional track lengths on the front or sides. My model is finished in Panzer Grey (Humbrol HM 4) with Dark Slate Grey random camouflage. Additional references can be found in Bellona Prints Series 29.

PanzerKampfwagen IV Ausf D

As a companion to the Standard Pz IV the, Ausf D as used in the Polish, French and North African campaigns is a simple enough conversion project. Study the drawings and note carefully the detail differences. Before starting construction some parts will require modification and the first of these are the two hull sides (parts 9 and 38). Carefully remove all but the first and the last suspension limit stops. Thin down the spokes of the rear idler wheels (parts 28/92 and 57/58). Add a 4 mm diameter disc of 30 thou plastic card to the sprocket hub (parts 27 and 56). Carve away the locations for the spare track on the front of the hull bottom (part 81). Remove the ventilator cowlings and the hinges from the final drive access panels on the hull front (part 79) and fit new Microstrip hinges. The spare wheel carrier (part 80) is not used so fill locating holes. Carefully carve away the engine cover grilles (part 78) but leave the hinges. Cut away the starboard hull top front (part 78) to create the stepped front and alter the side (part 75) to fit. Make a new machine-gun position front etc. Cut away the location for parts 83 and 88 on the hull rear. Make up a larger fuel tank to replace parts 86/87 and make new supports or use two parts 85 if you have them. A new exhaust system

The Sturmgeschutz IV (L48) is a combination of Stug III and PzKpfw IV as can be seen in this view of the unpainted model. The over-thick aerials seen here were replaced with stretched sprue.

Sturmgeschutz IV model finished in dark grey with slate grey mottle camouflage. Note new aerials from stretched sprue.

can be made from scrap and sprue to the shape shown on the drawings. The hull can be assembled as these alterations are made and then the final detail added. A new 'square' hull machine-gun mount replaces part 91 and a double shutter driver's visor in place of part 90. Fit the two towing points from the ends of the spare track (part 89). Fit headlights etc, and a channel to take the retracted aerial to the starboard side of the hull and a step to the port side trackguard, all as shown on the drawing.

The turret (use the F1 version parts) requires a few alterations. Cut off the corners of the gun mount (part 63) to the plan shape and of course drill out the barrel end (part 62). Carve away the ventilator at the front of the turret top (part 68) and fit an additional one further back as shown on the drawing. We shall not use the rear stowage box so the locating hole should be filled with body putty. The side access doors (parts 72 and 73) can be used for a PzKpfw III conversion as new ones have to be made as single doors hinged at the front

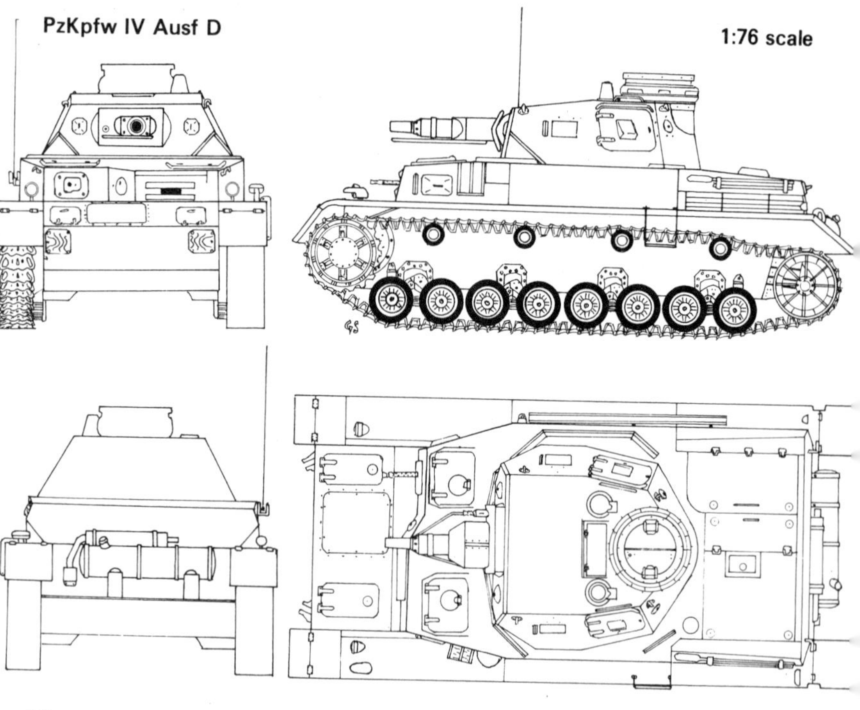

Above *the PzKpfw IVD (fore-ground) alongside a PzKpfw IIIF, which is another fairly simple conversion from the Airfix Stug III kit.*

Centre *more advanced PzKpfw IV model is this Ausf J with extensive additonal armour protection around turret and hull sides.*

Right *incorporating a completely new top on the PzKpfw IV chassis, the Hummel III/IV also uses parts from the Stug III and is only a conversion to be tackled when you have gained more experience.*

for the Ausf D. Fit the aerial deflector under the gun barrel.

My model is finished in early Panzer Grey and the crew come from various Airfield Service set personnel.

I hope that the conversions I have tried to describe will serve to introduce you to this fascinating hobby and that, after you have gained more experience, you will go on to do some more complicated 'scratch-built' models. There are many hundreds of different types of trucks, tanks, and specialist vehicles from which to choose and by keeping to this handy 1:76 scale a varied collection can be built up.

appendix

Basic book list

Suggested books that you may like to purchase or borrow from your local library. These are only some of those which I personally find most useful and to which I am constantly referring. The bibliography lists additional volumes which, though valuable to have to hand, relate more to specialised subjects.

Bellona Military Prints. Contain drawings, photographs and detailed technical information. Write for full list to: MAP Ltd, PO Box 35, Hemel Hempstead, HP1 1EE.

John Church. Drawings of mostly British, American and Commonwealth vehicles. Write for full list to: John Church, 'Honeywood', Middle Road, Tiptoe, Nr Lymington, Hants, SO4 0FX.

Len Morgan. Drawings of British, American, Commonwealth, Italian, German etc AFVs and soft-skin vehicles. Full list from: L.G. Morgan, 34 Midfield Court, Thurtlands II, Northampton.

MAFVA. Bi-monthly magazine *Tankette,* distributed to members, contains plans, photographs, technical and historical information, kit and book reviews etc. Secretary is G.E.G. Williams, 15 Berwick Ave. Heaton Mersey, Stockport, Cheshire, SK4 3AA.

Observer's Fighting Vehicles Directory, World War II by Bart H. Vanderveen. Warne.

British and American Tanks and World War II. Chamberlain and Ellis. Arms and Armour Press.

Russian Tanks 1900-1970. John Milsom. Arms and Armour Press.

Profile Publications. Various bound volumes of selected Profiles which are also available separately and deal in detail with selected subjects.

British Military Markings 1939-1945. Peter Hodges. Almark.

Wehrmacht Divisional Signs 1938-1945. Theodor Hartman. Almark.

Bibliography

German Combat Uniforms 1939-45. S.R. Gordon-Douglas. Almark.

Waffen SS. D.S.V. Fosten and R.J. Marrion. Almark.

US Army Uniforms 1939-45. Roy Dilley. Almark.

Japanese Army Uniforms and Equipment 1939-45. R. Dilley. Almark.

German Anti-tank Guns 1939-45. T.J. Gander. Almark.

Wehrmacht Camouflage and Markings 1939-45. W.J.K. Davies. Almark.

American Military Camouflage and Markings. T. Wise. Almark.

Halbkettenfahrzeuge: German half-track vehicles. J. Williamson. Almark.

Panzerjager. Chamberlain and Ellis. Almark.

Sherman Tank 1941-45. Ellis and Chamberlain. Almark.

American Armoured Cars 1940-45. Ellis and Chamberlain. Almark.

Soviet Combat Tanks. Ellis and Chamberlain. Almark.

Squadron/Signal Publications: *Panzer III in Action; Schutzenpanzerwagen in Action; German Half-tracks in Action; Panzerspahwagen in Action; Nashorn, Hummel, Brumbar in Action.*

A Source Book of Military Wheeled Vehicles. Edited by Bart H. Vanderveen. Ward Lock.

Military Transport of World War II. C. Ellis. Blandford.

Tanks and Other AFVs of the Blitzkrieg Era. B.T. White. Blandford.

Bellona Vehicle Data early series 2-view drawings only of vehicles, including photographs and technical details.

US Military Vehicles World War II.

Fighting Vehicles. Ellis and Chamberlain. Hamlyn.

Purnell History of the Second World War. Weapons Books (paperbacks): *Tank Force, Panzer Division, The Guns etc.*

Olyslager Auto Library, *The Jeep, Half-Tracks.* Warne.